LEARNING AB[illegible] WITH

Insects of the Southwest

Floyd Werner, Ph.D. & Carl Olson, M.S.

Illustrations by W. Eugene Hall

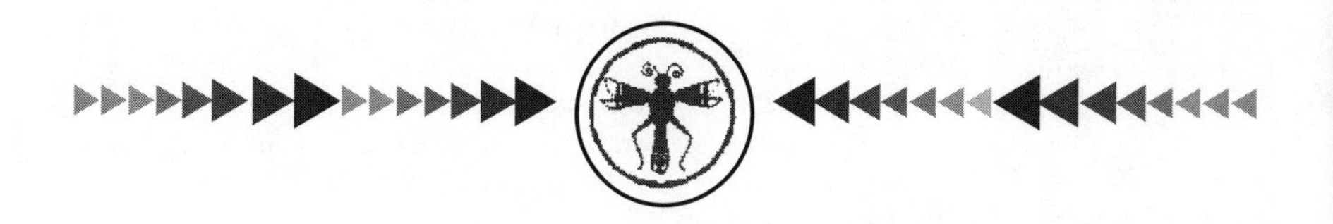

How to identify
helpful, harmful and venomous insects

Publishers: Helen V. Fisher
Howard W. Fisher
Fred W. Fisher
J. McCrary

Editor: Fred W. Fisher

Book Designer: Edgar H. Allard
Cover Designer: Paula K. Peterson

Published by Fisher Books, LLC

Fisher is a member of Perseus Books Group

Library of Congress Cataloging-in-Publication Data

Werner, Floyd G.
Insects of the Southwest: how to identify helpful, harmful, and venomous insects / Floyd Werner & Carl Olson; illustrations by W. Eugene Hall.
p. cm
Includes index.
ISBN 1-55561-060-9
1. Insect pests Southwest, New Identification. 2. Beneficial insects Southwest, New Identification.
I. Olson, Carl E. II. Title.

SB934.5.S68W47 1994
595′.2046′0979 dc20 94-29991
CIP

Printed in United States of America
Printing 10 9 8 7

Notice: The information in this book is true and complete to the best of our knowledge. It is offered with no guarantees on the part of the author or Fisher Books. The author and publisher disclaim all liability in connection with the use of this book.

Fisher Books are available at special quantity discounts for educational use. Special books, or book excerpts, can also be created to fit specific needs. For details please write or telephone.

Contents

Introduction

by Floyd G. Werner

The aim of this little book is to bring together a bit of information on the insects and other arthropods that people living in the desert Southwest are most likely to see and become interested in or concerned about. By no means is it a complete coverage. The University of Arizona insect collection has more than 13,000 identified species of only Arizona insects. There are many more that we have been unable to name or are waiting description.

The species included that are damaging in some way are evaluated to an extent, but no attempt has been made to include methods of control. For one thing, the chemicals that may be used keep changing; one that is in general use now may be removed from the marketplace next year. From a legal point of view, the best recommendation is to read the label if you use a chemical. The manufacturer has had to demonstrate its effectiveness (efficacy), and the recommended dosages have been adjusted to stay within a reasonable margin of safety.

For the non-damaging species, the tone is much more toward finding some reason for *not* doing anything, or actually *enjoying* the presence of the critters. Not everybody is willing to do this. I saw a grown man, outside and not even near his home, squash a wasp that was pulling a paralyzed cricket across the ground. He did this for no particular reason, except that he obviously didn't know or care what was going on.

In one's own patio, yard or garden, trying to keep the area free of insects and other arthropods is no small effort. I most often hear the fear expressed that small children *might* be in danger. So the place is kept "clean." Nobody asked the children.

I anticipate that some of the readers of this book will expect a formulary of the pests of house and garden, and how to control them. An earlier version of this manuscript did have some space devoted to control, but I have thought better of it. In this litigious society I expect I would have been sued by the vendors of the materials and methods not mentioned, and the long arm of the law would have been on me as soon as a pesticide that was legal at the time of issue got listed. When I received a contract from the publisher, Fisher Books, I found that I would be expected to stand alone in any suit brought against the publisher or myself. Thus, no chemical controls are presented.

The best advice for people who want to use pesticides is to look at the selection available where they do business and follow the label directions and restrictions exactly. A great deal of effort has gone into testing all of the available materials. To be registered for use they must be proven effective at the concentrations given on the label. And they must be safe to use, with a good margin of safety. The USDA and the EPA have been interacting for years to produce the labels.

Instead of giving you ways to eliminate, rid or destroy these animals, I want to do what I can to allay the fears of the citizenry, fears that in many cases are completely unfounded. I even have to be careful here, because I don't want to trap someone into a situation where some beast that was supposed to be harmless turned out to cause some damage. I'll content myself with telling the reader how we treat the creatures when we encounter them in our own home or yard, or out in the desert or forest.

I particularly want to help the youngsters who first discover these little creatures to be free to see and enjoy them. A paranoid parent can scare the mystery and delight right out of the transaction. Perhaps an over-instructive parent could be as damaging. I have put in little notes about natural history and behavior, more to get the observer started than to provide all the answers. One thing that initiates observation in young entomologists is discovering that too little is known about even our common insects near the house. A careful observer can be right at the boundary of knowledge using only the unaided senses and never going more than a few yards from home.

A dismaying aspect of the life of an entomologist is that everybody expects such a person to know all about how to "control" insects. My interest in insects started early, and with butterflies, as

is often the case. By the time I was in high school my interest was well known in the neighborhood, and I had to make a wide detour around one house in particular if I wanted to go anywhere during the growing season. The man of the house had a little garden, with one or two plants of most vegetables, and a few flowers. He invariably nailed me with the inevitable question "How do I get rid of these bugs?"

Years, military service, a full-blown scientific expedition and a couple of college degrees later, I ended up at the University of Arizona. Here I found a situation in which everybody in town seemed to expect the "experts" to drop everything and make a quick trip to look at some "bugs" destroying their plants. The equivalent of my nemesis in high school had a short row of about every crop that would grow in this climate.

My background is as a naturalist, and my continuing interest is in the development and use of classification of insects. Diversity fascinates me. The Southwest has a concentration of diversity that is unbeatable in the United States. So this has become my chosen home.

Please don't think me callous. There are some real problems with insects and other arthropods, and people need help in finding answers to the problems. Part of the power of a land-grant university is to identify these problems and find solutions. What seems to have been an inevitable consequence is that now everybody seems to think that the wonders of chemistry need to be called on to handle even the imaginary problems.

Somehow humans have been trained to consider anything that walks, flies or crawls near us as something to eliminate, whether it is affecting anything of value to us or not. Just go to the supermarket and get some spray and take care of the problem.

I have a belief that the fear and revulsion reactions are learned, despite the best efforts of many in the educational system to encourage the interest children have in the world around them. Birds fare well in the esteem of the populace, although a few people, such as my wife, are terrified at a close encounter (conditioned response from a rooster that used bluff to gain cookies). Most mammals also fare well, if they are not too large or too small. They have such soulful eyes. Reptiles don't do too well. Despite much TV time about the value of snakes in keeping down rodent populations, the last words of the anchor folk let the audience know that they don't want to have anything to do with snakes, venomous or not.

When it comes to the creatures that have a body design very different from ours, empathy seems to diminish. In addition, the reputation that some of them are venomous reduces their circle of friends. The child who might really like to have a pet spider or a grasshopper might be allowed to have the grasshopper, but the spider is venomous so "no" is the answer. Even worse is the reaction of a parent who has been conditioned to cringe in terror. It is hard for a teacher to overcome learned behavior in the child. Even worse, the teacher shows the same reaction.

I don't expect everybody to be enthusiastic about having the house crawling with bugs. I even admit that there are some that I don't tolerate. But I like to think that I can give a logical explanation for the ones I can't stand. Maybe I have been conditioned too. Pesky house flies, eye gnats, fruit flies, mosquitoes, most species of cockroaches, and black widow spiders are not welcome. The cockroaches really are quite innocuous, keeping out of sight during the day. Many have tried to count them as filthy, but the evidence that they can spread bacteria harmful to mankind is slim. I was always able to get a reaction from the young ladies in my classes by observing that monkeys seem to have learned about the defensive scent glands on cockroaches and consider them unfit to eat. Maybe that is the real reason that our species has such an aversion to them. If I lived in a desert foothill area where they occur, as many people do, brown spiders, scorpions and cone-nose bugs would also be most unwelcome.

The purpose of this little book is to introduce the reader to a fascinating assemblage of creatures which share our habitat. Not all are always completely welcome, but some are, and most can be tolerated with little effect on our well-being. Some are venomous, and these are all mentioned. People are concerned about them, particularly when there are small children in the home.

Maybe I can induce parents to encourage their children to be curious about the little creatures around and not just tolerate them. Learn more about them, play with them. By this I don't necessarily mean going to the library for an armful of books to do the learning. This is a possibility, but the task is most difficult. There is such diversity that it takes a carefully selected 10-foot-plus shelf of books to even get started. The huge diversity forces one into the unfamiliar

world of Greek and Latin scientific names. Just pronouncing the name may be a challenge. Learn by observing. Learn by keeping them as pets. If that caterpillar that was supposed to become a butterfly or a moth according to what you learned sprouts a fly or a wasp, that is the time to hit the books. You have discovered parasitoids, important creatures in the natural regulation of insect populations in the wild. Maybe you are witnessing the start of a budding career in the field of biological control of insects.

Don't expect everything to turn out educational though. Maybe tethering a fig beetle and having it fly in circles will teach something about flight. Mostly it is just fun, with no harm to the beetle. It can be used to teach compassion by having the beast released when the fun is over. It may be too gruesome for the parent to watch, but there are predators galore in the small-size range, and watching them in action matches the most gruesome acts on TV. I remember well an entomological home movie of a praying mantis eating a caterpillar, alive of course. A few appropriate sound effects would have made it better. The charm is that these are things that are happening in the real world, and the observer isn't being influenced by the dulcet tones of the TV host telling you what you are seeing.

The first section treats the insects and other arthropods that may be found in and around the home. These are covered in detail. Not all will be present in every homestead, but a large percentage will be at one time or another. Some of the household insects are real invaders, causing considerable annoyance or even damage. Others are benign and can be considered more as *house guests.* I hope that a fuller knowledge of what the individual species do for a living will make the reader more tolerant.

Acknowledgments

I would like to extend thanks to Renée Lizotte, Mike Lindsay, Gene Hall and David Collins who read and gave very thoughtful suggestions to the improvement of this book. I also want to thank the University of Arizona Department of Entomology for their support in the development and preparation of this book, especially Dr. Elizabeth Bernays, who spurred Floyd into starting this book and who later put Fisher Books and the authors together.

—Carl Olson

So Naturalists observe a flea
Has smaller fleas that on him prey
And these have smaller still to bite em
And so proceed ad infinitum.

Jonathan Swift
(Poetry, Rhapsody, 1733)

The Bugman's Philosophy

by Carl Olson

Working with insects has been a love affair of mine since as long as I can remember. As a small boy with the countryside as home, I first collected butterflies. Then I learned that it was more fun to collect caterpillars and learn firsthand how they developed. Consequently, our garage was filled with jars containing all imaginable kinds of caterpillars, from Io caterpillars with their urticating spines to enormous green Polyphemus caterpillars with their startling array of colored tubercles.

When I finally got to the University of Arizona and started working with Floyd Werner, I had become indoctrinated into aquatics, not as the fisherman knows them, but as a biologist studies them. The joys and frustrations of capturing dragonflies are to me as stirring a hunt as stalking an elk, and most of the time just as difficult. I had experienced in graduate school what most desert dwellers dream of, wading in the water while enjoying the outdoors, but I also had a good idea of what kind of exciting creatures I would find beneath the water hiding perhaps under a rock.

My real awakening to the world of insects came by being in the field with Floyd, who unlocked many more doors of entomology with his keen eye and pioneer spirit. We spent hours exploring and discovering what the desert Southwest has waiting for the curious. Now I hope to carry on that naturalist-type position that Floyd epitomized until his death in 1992. I had the best of teachers to help me develop a broad knowledge of this fascinating microworld, and now I hope to share these joys, too. I am the preacher of good news when it comes to insects and what they do for us. I learned early to have confidence in insects and believe in Aldo Leopold's land-ethic philosophy, which sums up for me what bugs really do on this planet. They may take, but in most cases they give back to the system to keep it healthy.

I do not find nature's creatures to be selfish. On the contrary, they seem to have that built-in guide to stop them from messing up a good thing. It is the aim of most living things to maintain diversity, and this will then keep a healthy ecosystem. Once greed and selfishness creep into a system, you can bet trouble will soon follow. With insects I seldom find this to be the case. Oh, to the human's aesthetic taste, insects are the destroyers, but to Nature, they are the recyclers, the reworkers and the designers. They've been at this game of survival much longer than human beings, and they've surely succeeded.

As I wander about this desert, I now try to figure out what insects are around making this world so delightful to me. I may become an oasis in the desert to the creatures. If that fly comes to my nose and becomes an irritation, I may shoo it away but I know why it comes to me. Life won't always be as I wish it to be, so maybe I can return a little of me through some other life forms. Peace of mind comes in many ways, but to me it comes in the form of diverse shapes and sizes of all those microcreatures surrounding me in life.

The writing of this book seems to have taken ages as I look back. Floyd Werner began the task in 1989 after he retired. We discussed what insects, etc. would be good to include and then he sat down and wrote. During the three years Floyd worked on the book, many obstacles slowed its progress as Floyd battled ill health.

Finally in November 1992, Floyd asked me to complete this book, as he was one to never let things go unfinished. With great pride, I took the manuscript and began adding my touches. Needless to say, this project is a labor of love, both for Floyd and for the world of insects. This book writing is quite a job.

As you read the stories, they reflect experiences of both Floyd and myself. You may well detect two writing styles and hopefully they will window our souls to you. Some you will easily recognize as Floyd's touch, because of the time factor. Others, well just guess!

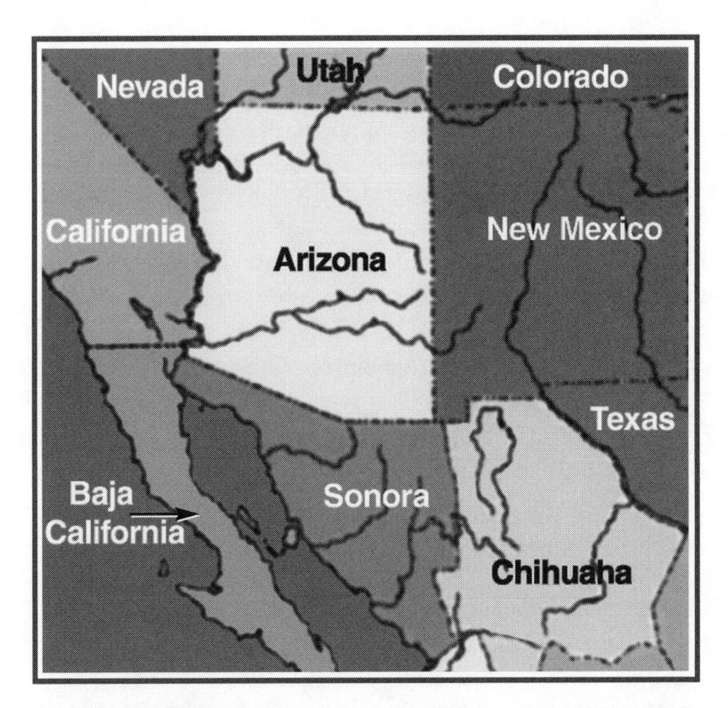

As the reader homes in on a region of interest, it is good to have a feeling for the geographic boundaries. When studying insects, because of their importance to humans and the natural world, it is important to see where they live and which ones might be encountered during travels or home changes. Because of the mobility of insects, this map by no means indicates a rigid distribution. Walls cannot be built high enough to stop the bugs, but the map is very representative of the home range of the critters.

Chemical Controls
A Word of Caution

Chemical control is mentioned only in that chemical control may be necessary. The interplay of chemical companies, advertising agencies and governmental regulation keeps a steady stream of new materials on the shelves of supermarkets all over the country. Each has been developed for specific uses, and safety has been built into the regulations printed on the box or bottle. If all precautions are observed, there is a wide margin of safety. Don't use anything without reading the label carefully and observing all directions. One Tucson lady decided to free her house of pests with aerosol bombs. She opened several, closed up the house, and watched the effects of a real bombing. She hadn't turned off pilot lights, as the printed precautions had insisted. One door left the house in its frame.

My aunt went after a cockroach with an entire can of aerosol spray, leaving the poor roach drowning in poison, a bit of overkill. After removing the dead roach, she didn't bother to clean up the floor. Later, coming down the stairs, she slipped on the residue and broke her hip. Misuse of pesticides results in many unfavorable situations, so use them judiciously if you must use them at all.

A Word about Insect Measurements

A Bugman describes an insect's size by giving the length of its body (usually excluding wings and legs). In this book the insects are frequently shown two or four times greater than their actual size. In these instances, a "bracket" is placed next to the insect to indicate the actual size of the body. This example shows the Dermestid Beetle Larva with a bracket measuring 1/4" (representing the actual body size) although the illustration is approximately three times the actual size.

CHAPTER 1 House Pets and Pests

The insects and spiders that live inside the house are well beyond the limit of tolerance (zero) of many people. Yet they are just as interesting as their kin outdoors. Most of the spiders that live with us are more pets than pests. Perhaps the most fastidious Puritan housekeeper would not agree. For the cost of a few strands of silk we get interesting creatures that actually make a living eating some of the insects we call *pests.*

Encouraging the pets takes little more than avoiding the use of residual insecticides and holding back on the impulse to smack or step on them. Night lights are particularly useful to the spiders, which can set up housekeeping near them and let the food supply pour in.

Some insects fit into the same category. Others are either so annoying or so repulsive in their habits that most people regard them as pests. Cockroaches are favored by few, and houseflies by even fewer. I have suggested a tolerance level for many of the examples, biased, of course, but perhaps useful as a point of departure.

The key to managing pests is to keep them out of the house in the first place, and to keep from providing them with a ready food supply. If all crumbs are swept up, there is little for a cockroach to eat. Cockroaches and crickets have a severe water problem so they make frequent nocturnal trips to sinks and are often trapped by the smooth porcelain walls of the sink. Sticky-walled traps placed near the sink are very effective in reducing their numbers. Adult and larval beetles in stored-grain products, dried fruits and nuts can be handled by placing such products in tight cans or jars. If isolated products do become infested, the creatures can be killed by placing the container in a warm oven for several hours or in the freezer for a day or so. Repeated infestation usually means that you are buying infested foodstuffs.

COCKROACHES
(Order Blattaria)

Cockroaches have many features in common. Their body design makes it possible for them to squeeze themselves through incredibly narrow cracks. They all have some odor and they all lay eggs in egg cases, known as *oothecae,* either placed in their habitat or carried around by the mother. All of the species found inside houses here have a critical dependence on a water supply.

In teaching beginning entomology courses I could always get a rise from the class by suggesting that the housewife's abhorrence of cockroaches is inherited from ancestors in the dim and distant past, who discovered that they were quite inedible because of their defensive secretions.

Sealing holes around plumbing fixtures and between apartments, using sticky traps, and practicing general cleanliness in the kitchen can usually be relied on to keep these pests in check.

Encounters with roaches, especially sewer roaches, occur when the weather becomes very extreme in the Southwest. If the hot dry months really become extreme with many days over 100F, you'll begin seeing them indoors. When the summer monsoons hit in July-August, again the numbers of roaches coming indoors increases. These movements are due to the disruption of their normal habitat, and the unlikely but always present invitation to nicer haunts, in other words our homes.

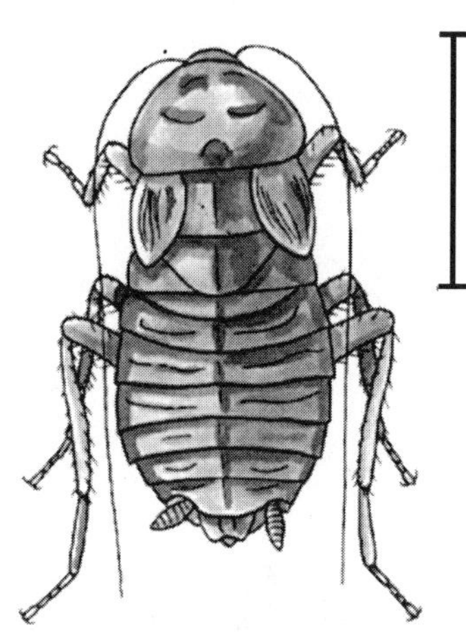

Oriental Cockroaches
Twice actual size, male above, female below

AMERICAN COCKROACH

(Periplaneta americana)

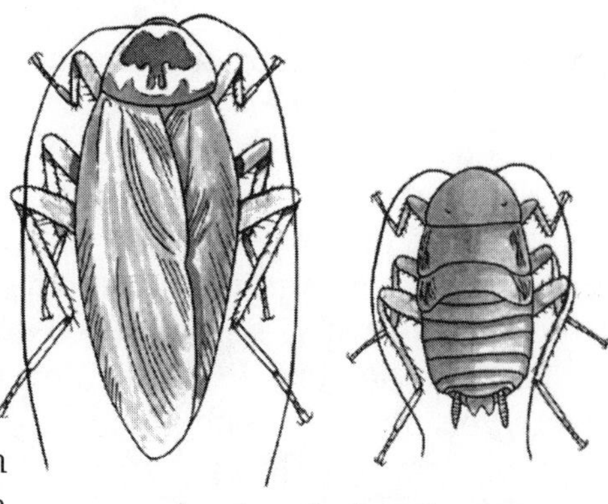

American Cockroaches
Actual size: adult left and nymph right.

This is the cockroach that most people refer to when they tell about the rodent-sized varmints they saw in Texas or Florida. There are much bigger species in Panama and other parts of the Tropics. Adults are almost two-inches long, males almost identical with females, shiny and brown with some tan markings.

Adults have long wings, and are sometimes seen in flight. Nymphs are rarely seen around houses. The typical place for a thriving colony is inside the sewer system of a city. Egg cases are deposited in cracks in the walls and all stages of nymphs and adults live above the water in the system.

Control of cockroach populations in the U.S. is a municipal function. It involves coating the walls of the main chambers under the manholes with a persistent pesticide. Some cities do this on the basis of citizen complaints, but a scheduled application through the whole system has to be more effective. A strange black wasp is often seen in areas where this roach is abundant. This wasp, an evaniid, has a round abdomen attached to a long thin petiole. One notices this wasp because it waves the abdomen up and down like a flag when it alights on the window sill or other perch. This wasp parasitizes the roach eggs and helps manage the roach population.

Other reliably moist habitats may occasionally hold populations as well. These may range from concrete-block walls, large holes in trees with rotten wood, to the debris that accumulates at the bases of palm fronds. These sources are much more minor than the sewer system, at least in a desert environment.

Adults that get into houses are most often seen associated with sinks or bathtubs. It is most likely that they had been seeking water to drink and gotten trapped by the steep, slippery sides. The common belief is that they have made their ways through the drain traps. This has been demonstrated as a possibility, but is probably not the usual route. Either way, they are probably *not* welcome.

This was probably an African species at first, but it seems to have gotten scattered around the world in sailing-vessel days. It got its scientific name, *Periplaneta americana,* from specimens collected in North America and the common name was manufactured from the scientific name. Before modern sanitation methods, huge numbers bred in dumps in cities of all sizes. The species is more tolerant of lower temperatures than any of the others that we see regularly. There seem to be no friends of the American cockroach. Its candidacy for nomination as the official U.S. insect was resoundingly defeated by the membership of the Entomological Society of America, in favor of the monarch butterfly.

BROWN-BANDED/GERMAN COCKROACHES

(Supella longipalpa)/(Blattella germanica)

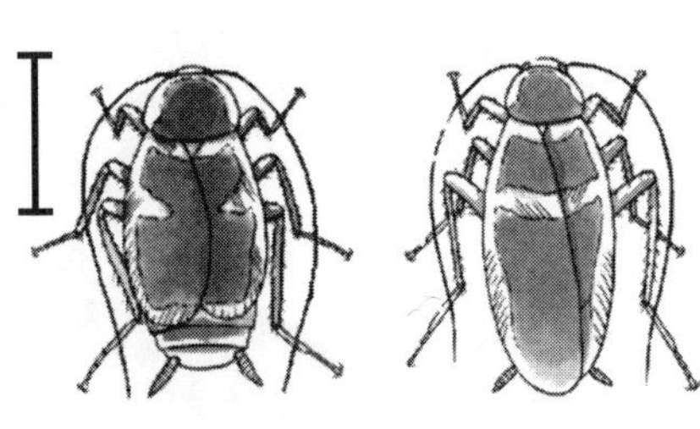

Brown-Banded Cockroaches
Twice actual size: female left, male right.

These are the only cockroach species that are likely to establish permanent residence in houses in this region. The male brown-banded is narrow with long wings. The female is pear-shaped bearing short nonfunctional wings. All stages are shiny, tan, banded in a chestnut color that would be attractive if people could tolerate roaches. The German roaches are recognized by the short black stripes worn on the shield directly behind the head. Both males and females can fly, but are more likely observed scurrying away on the countertop when a light is turned on late at night. The most likely place for a population is the kitchen, where there is a ready supply of food and access to water at the sink. The brown-banded roach tends to hide in cabinets, drawers or behind pictures on the wall, whereas the German insidiously hides inside electrical

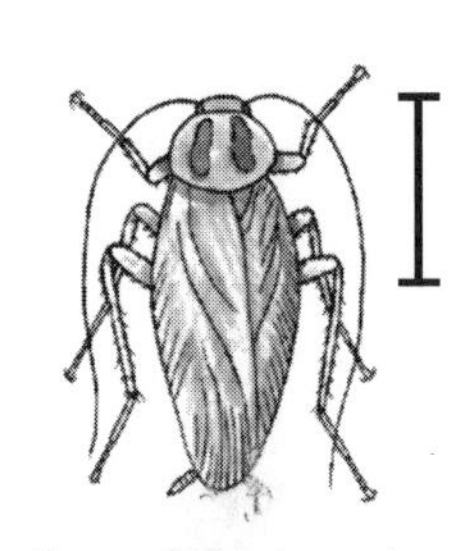

German Cockroach
Twice actual size

appliances, making control most difficult. When the female is ready to reproduce, she deposits her egg cases in hidden places, often along corners and edges of the undersides of counters and drawers. The egg cases are distinctive, thick bean-shaped, and less than a quarter inch long.

Cockroaches are strictly nocturnal in their activities, so may be very abundant and not be noticed during the day. Reproduction is rapid, so a few months of neglect can lead to a population explosion. In apartment buildings the most likely source is probably other apartments, with access from room to room through holes for pipes. In isolated homes used furniture can be a prime source, or anything brought in that has been in an infested place elsewhere, such as corrugated cardboard boxes. Elimination of the problem sometimes requires drastic action, with sanitation an important factor.

These species do not live out-of-doors here, and are never involved in cockroach populations in the sewer system. They can live in drier places than the other species, so long as there is a tiny amount of water available somewhere. Generally regarded as pests, with which I agree. They really aren't at all filthy, though. The notion of being disease carriers has resulted from a few restricted occurrences.

TURKESTAN COCKROACH

(Blatta lateralis)

This species is a recent introduction in the country that is spreading rapidly in the United States. Adults can be an inch long. The males are fairly slender with long, yellowish wings. Females are broader with very short, rounded nonfunctional wings. The female is dark brown to black, decorated with creamy markings on the shield and a short cream-colored stripe edging the shortened wings. Nymphs are wingless, chestnut brown in front and black on the rear.

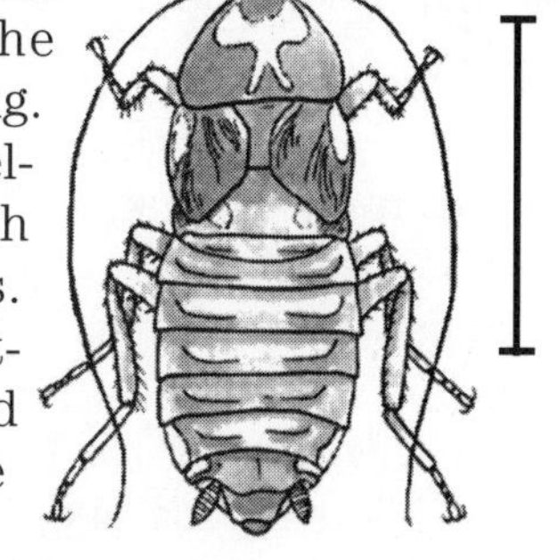

Female Turkestan Cockroach
One and one half actual size

Compost piles and bags of mulch can house very large populations. Egg

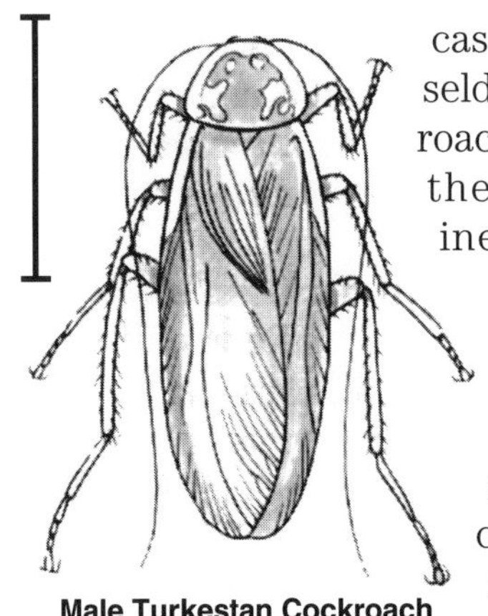
Male Turkestan Cockroach
Twice actual size

cases are deposited in such places, but are seldom seen. The occasional Turkestan cockroach that gets into a house behaves in much the same way as other roaches, almost inevitably seeking out the sink or other source of drinking water. Males, because of their flight ability and attraction to lights, are the more commonly encountered ones indoors. Because these don't start massive colonies and are quite clean, the horror of discovery doesn't require the intensive screams that other roaches might elicit. Because their main habitat is out-of-doors the citizen might assume that control would be a municipal problem as it is for the American cockroach. This really is not the case. Anyone who has large numbers is almost certainly generating them on his own property. Trapping or killing those that get into the house is more feasible than attempting to eliminate all sources outside. I have noticed the gecko population on the rise since this roach became established in my yard. Because its major habitat is out-of-doors, this species would fall into the tolerable category if it weren't a cockroach.

DERMESTID BEETLES

(Family Dermestidae)

Dermestid beetles are very likely to be found in stored products, but adults and larvae are very different from the beetles listed under weevils. Adults are oval, about an eighth inch long, and dark with a pale pattern. The larvae are slender, tan, about a quarter of an inch long at largest, and very hairy with a tail at the hind end.

Dermestid Beetle Larva
Three times actual size.

While they may be found in stored grains, these beetles and their larvae are more likely to be found in many other products, particularly those that have a high nutritional value. Nutmeats are favored. Because of this, the search for centers of infestation must be wide if one attempts to control dermestids. Dry pet food is a very common source, because it is not stored as carefully as human food. The strangest source of infestation I have discovered was in d-CON® rat poison. The bait portion, consisting of rolled oats, was a delightful meal for the dermestid larvae and the poison part of this product had no effect on the blood of these insects.

Finding the source is hampered by the fact that the adults are inclined to fly to light and may accumulate on windowsills, far from the site of infestation. They may even lay eggs and start a small infestation on dead insects in such places. Light fixtures trap them, and fluorescent fixtures even remain cool enough to serve as sources of infestation. The need to find the source is enhanced by the fact that the larvae may feed on woolen and silk clothing in the same way as the more infamous clothes moths.

Clothing stored without being cleaned is particularly susceptible. The number of larvae present is rarely very large, but even a few round holes can be very expensive. Clothes moths cause similar damage, distinguishable by the presence of silk strands laid down by the larvae, but these are rarely seen in this region. A close relative is the carpet beetle, which can ravage the nap before one is aware of their presence.

FLIES

(Order Diptera)

Flies appear to be the universal nuisance to human beings. They seem to be everywhere, always landing on one's face or arms. When no other insect seems to thrive, there is the fly. The question that always follows then is "What good are flies?" To answer it, one must look at the life history of flies.

The juvenile stages of flies, termed *maggots,* live in situations that most other insects would disdain. Maggots feed on everything from fresh manure to decaying corpses, acting as the ultimate recyclers of Nature. Because all these situations are readily available in every place that has living beings, flies will always be with us, and doing beneficial things for us.

Adult flies also contribute a lot to life. One may notice what at first appears to be a bee or wasp hovering or landing on a flower. A closer inspection may well reveal the insect to be a fly. Many adult flies, with their hairy bodies, are perfect pollinators. Adult flies may also search out other types of insects to lay eggs on. The maggots that emerge then feed and kill this host, thus acting as a population-control factor. The term *parasite* usually makes a person shudder, but here is a positive use of parasitism.

Sanitation is the best way to keep fly populations away from houses. Cleaning up garbage, animal droppings or other similar food sources will prevent flies from developing. No food is a great deterrent. If populations of adult flies do enter the house and one doesn't want to put the fly swatter into action, a pan with a bit of detergent should attract and disable them.

Why do flies alight on one's face or arms? In most cases, the flies have come because of the moisture. All insects need water, and the sweat on people is a perfect oasis for the flies.

BLOW FLIES
(Family Calliphoridae)

Blow flies, also called *bottle flies,* are the large, shiny green or blue flies that get into houses and make a nuisance of themselves by buzzing from window to window. Screening keeps them out, of course, but a few make it inside. They are larger than houseflies, definitely noisier, and usually more adept at avoiding the swatter.

The larvae of these flies are associated with dead vertebrate animals. Females are attracted to an animal soon after its death, and lay masses of eggs around the face or other openings. The larvae that hatch from the eggs eat their way into the carcass. The name of the flies comes from the observation that the larvae are on the carcass by the time it is swollen, or *blown.*

Because the larvae are among the insects that quickly convert a dead animal to compost during hot weather, they have to be considered as useful. As the science of forensic entomology grows, these flies will gain more respect as these are the tellers of tales that help solve death mysteries.

FRUIT FLIES
(Family Drosophilidae)

The nuisance insect that goes under this name is not related to the famous Mediterranean fruit fly, but belongs to another fly family,

Drosophilidae. It is the fruit fly of the geneticist, *Drosophila melanogaster,* the species most common in houses. *Drosophila funebris* is a darker species associated with compost and garbage out-of-doors.

The fly itself is less than an eighth inch long, tan with a black-tipped abdomen and red eyes. The larvae that go with these adults develop in the flesh of decaying fruit. Fruit flies have few friends. They are a nuisance around the salad bar, hovering and landing, particularly if salad dressing has been applied. In my home we have found that a particular brand of bread-and-butter pickles attracts them. An open jar drowns the lot. Persistent infestations are usually the result of sloppiness at the greengrocer's. Changing sources will usually solve the problem at home.

In the early days of *Drosophila* genetics the breeding chambers were milk bottles charged with ripe bananas. A tale from these early days at Harvard involves the refusal of the milkman to deliver any more milk until his bottles were returned. One of the graduate students, a founder of this branch of genetics, had almost filled the basement with milk bottles in which he was rearing his experimental animals.

HOUSE FLIES

(Musca domestica)

Screening and replacement of the horse as a universal means of transportation have cut down house fly problems in all parts of the developed world. But the insect is still with us. The familiar gray-and-yellowish fly is attracted to the odor of food, and can become particularly annoying at outdoor barbecues.

The larvae or maggots live in and feed on manure or decaying plant material of almost any kind so long as it is not too wet or too dry. In warm weather a generation takes less than two weeks.

Composting manure or scattering it so that it will dry rapidly usually takes care of the fly-breeding problem, but continued inoculation with parasitic wasps that feed on the larvae is often practiced where large numbers of animals are kept. In the very early days of DDT, applications of this insecticide wherever adult flies congregated had a dramatic effect on the flies, but house fly populations the world over soon developed resistance to DDT.

Adults have sponging-type mouthparts. They suck up liquid food or may regurgitate liquid to dissolve solids such as sugar. This habit, and their inclination to alternate between food and manure in their daily activity, has made house flies suspect in contamination of food with bacteria detrimental to man. Watch as a fly lands on a food source and you will see its mouth extend immediately. Flies taste with their feet and know when food is available in this manner. Incidentally, if what looks like a house fly lands on a person and inflicts a sharp bite, the fly is the stable fly, discussed elsewhere. I have very low tolerance for house flies landing on food, and suspect that few do tolerate them.

MOTH FLIES

(Psychoda spp.)

Tiny, delicate gray insects in the bathroom, holding their wings like a peaked roof and moving around with a stiff-legged gait, are moth flies. These are true flies but picked up this name because of their moth-like appearance and the fact that the whole body and wings are covered with long hairs that look almost like the scales on moth and butterfly wings. Moth flies are harmless, but even tolerant housekeepers don't usually welcome insects in the bathroom. They can be difficult to eliminate once established.

The source is the sludge that develops on the walls of the overflow drain where the larvae live. Soapy water splashing on the walls provides both habitat and food. Even if the water used is too hot for them to tolerate, the overflow provides a sanctuary. The larval habitat has to be cleaned out or subjected to enough heat to kill the larvae for the source to be eliminated. Nobody seems to have much good to say about moth flies. They are quite harmless but not very aesthetic as their other moniker, *toilet fly,* might suggest.

INDIAN HOUSE CRICKET

(Gryllodes supplicans)

About 1985 pale-brown crickets began appearing in houses all over southern Arizona and southern California. These are of about the size of the more familiar field crickets, perhaps a bit more slender, and paler than all but the palest field crickets. There seems to have been a

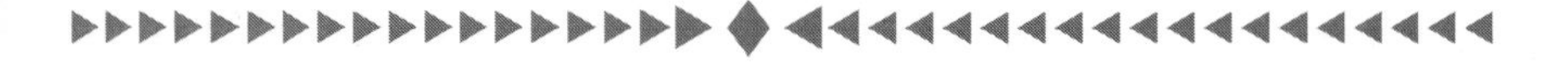

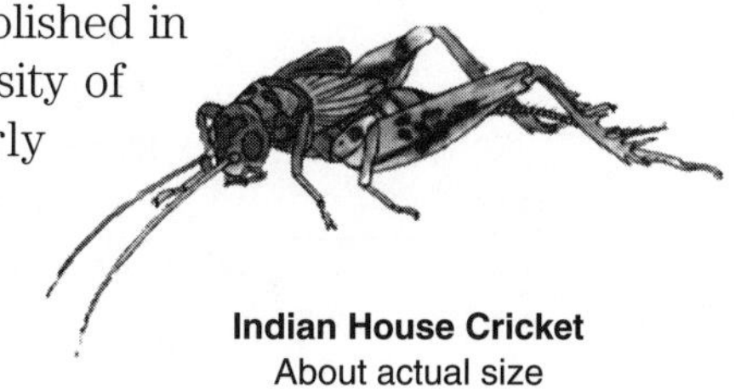

Indian House Cricket
About actual size

population of these insects established in the utility tunnels of the University of Arizona since at least the early 1950s, and probably elsewhere in such sheltered habitats. This species is not very tolerant of cold, freezing weather killing all that are exposed to it. One warm winter the crickets spread. Perhaps there was a slight increase in their tolerance to low temperatures at the time. At any rate, there now are Indian house crickets all over the place.

These crickets have some very definite preferences in habitat. As crickets they are largely nocturnal, hiding away during the day. Their resting place preference is strongly for vertical surfaces close together. Stacked flower pots are ideal. Clapboard siding or piles of boards leaning against a wall are almost as good. Indoors the population is maintained by the females laying eggs in the soil of flower pots. A tell-tale sign of these crickets is an incredible amount of small, black droppings covering the top of a box in your storage room.

Crickets have a wide range of food choices, including both crumbs of people food and tender plants. Seedlings may be eaten completely. The biggest objection to them is that the males have a very high-pitched chirp, which can be very annoying to those who can hear it. I am one of the fortunate people who do not hear in that high range. A singing male in a cooler duct is particularly maddening when one is trying to fall asleep.

Crickets already in the house can be controlled with sticky traps. Sealing all cracks in walls will help keep them out. Out-of-doors management probably is most effective when the preferred resting places are eliminated.

SILVERFISH\FIREBRAT

(Lepisma saccharina)/(Thermobia domestica)

The group of insects to which the silverfish and firebrat belong is one of the most primitive, tracing its ancestry back to a time when no insects had wings nor the ability to fly. Silverfish are rounded in front and tapered gradually to a three-tailed posterior. Adults are about half an inch long, and young look exactly like small adults. The body

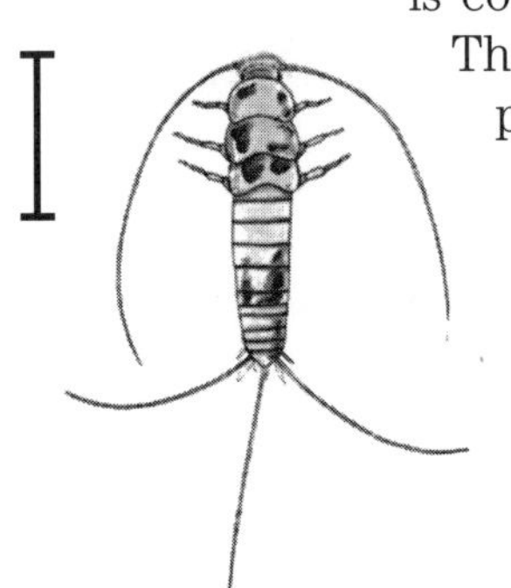
Silverfish
Two times actual size.

is covered with extremely delicate silvery scales. The firebrat's scales produce a tan-and-brown pattern. The silverfish is very widespread, and is quite likely to be mentioned on a pesticide spray can.

These insects feed on starchy materials, including such things as the sizing on coated paper and book bindings. Their mouthparts are not very efficient, and their feeding usually results only in the shininess of the finish being removed. Both firebrats and silverfish are found throughout the world. The firebrat got its name in England, from its inclination to move in close to the hearth. It has a remarkable tolerance to high temperatures, and is unable to reproduce at temperatures much lower than 100F.

SPIDERS
(Order Araneae)

The presence of four pairs of legs and a body made up of a cephalothorax (head and thorax combined) and an abdomen marks an arthropod as a spider. One to four pairs of spinnerets, small finger-like structures at the tip of the abdomen are the silk-spinning devices spiders are famous for.

All spiders have a pair of fangs up front, and most use the fangs to inject poison when they bite. The fangs are so small and the venom so weak in most spiders that they pose no threat to us at all. Yet spiders rank high on the hit list of many a householder.

GIANT CRAB SPIDER
(Olios giganteus)

The Giant Crab Spider is by far the largest spider found in houses. A fully grown individual measures two and a half inches across its long legs. The Entomology Department receives a steady supply of these

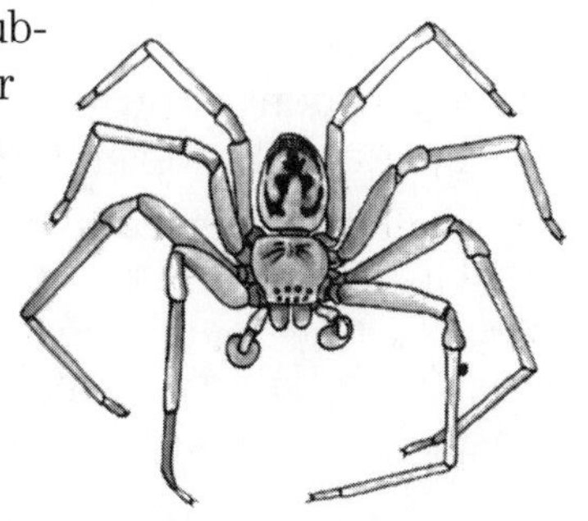

Giant Crab Spider
Actual size.

in jars, especially after there has been publicity about brown spiders. The popular press points out that a brown spider can be identified by a brown, fiddle-shaped marking on its back, a feature of many spiders including the giant crab. All kinds of brown spiders are considerably smaller than an adult giant crab. Most people don't want to look that closely, but the giant crab has four pairs of simple eyes.

These spiders wander about the house, leaving a trail of single strands of silk, which accumulates dust and contributes to the familiar "drapes" of poor housekeeping. Normally one notices giant crabs high up on the walls. When startled, it sidles away in a manner similar to the crabs on the seashore.

I would once have said that giant crab spiders are completely harmless. A young male tried to bite me when I had him cupped in my hands for a trip outside, but was unable to break the skin. However, a close friend did get bitten by an adult female, and felt pain for a full day. She no longer tolerates this spider at all.

It is remarkable that such a large predator can make a living indoors. Indian house crickets must provide a reliable food supply, so the person who is particularly annoyed by the chirps of male house crickets might do well to find a giant crab spider outside and move it in. We almost always have at least one in the house, although a visiting dog is hard on them. We regard them as pets, a designation not always shared by human guests.

PATCH SPIDER

(Family Oecobiidae)

Patch Spider
Actual size.

This spider is much more conspicuous for where it lives than what it looks like. Its shelters are round, slightly bulging patches about half an inch in diameter, placed in corners or hollows on walls.

Because they are constructed of slightly sticky silk, they catch a certain amount of dust. Most of these patches are devoid of life at any one time. However, if one keeps tapping patches one at a time, the small pale spider will appear and scurry to the next patch.

This creature has traveled around the world with man to the extent that its native home cannot now be determined. It is harmless, and makes a contribution by catching small insects. The vacuum cleaner is its worst enemy. I call this species a *pet.*

WEB SHAKER SPIDER
(Family Uloboridae)

The web shaker is the only spider that makes the traditional wagon-wheel kind of web inside houses. Out-of-doors there may be a much larger orb weaver spider that makes webs up to two feet across and guy wires beyond. The web shaker makes a small web, perhaps up to six inches across, and often almost horizontal. The spider itself is small, adult females being about a quarter inch long.

The spider gets the name used here from its behavior of bouncing the web, particularly when disturbed. The purpose of the bouncing almost certainly is to increase the volume of the space in which the web can entangle small insect prey. In the outdoor orb weaver the rays between the spokes of the web are sticky. The spider moves around holding the rays. In the web shaker the rays are made up of a tangle of fine silk strands, not sticky but tangled enough to catch the spines and hairs of the prey. These spiders, known as *cribellate spiders,* produce non-sticky silk from a group of tiny spinnerets on the abdomen and tangle it up with special combs on the hind feet. The remarkable wagon-wheel design of the web has been evolved independently in the sticky-silk and tangled-silk groups. Examination with a hand lens will reveal the web structure.

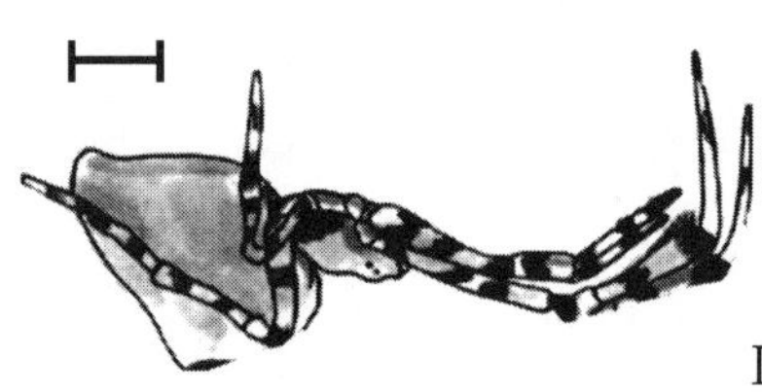

Hackle-web Spider
Shown eight times actual size.
Abdomen at left (shaded area).

No need to worry about being bitten because this spider is one of very few that lack venom. Opinions may vary on this one, pet if you don't have to keep things looking tidy, a nuisance if you do.

These spiders are also known by such names as *hackle-web* and *feather-legged* spider.

WEBSPINNER

(Oligotoma nigra)

Webspinner
Eight times actual size.

These insects would not be conspicuous but for the fact that the males are attracted to light and are slender enough to walk through most window screening. They are more than a quarter inch long, very slender, narrow-winged, and inclined to fly and land on anything near a light, including the face of a reader. They were once more of a nuisance than they are now, when we depended more on opening windows to keep ourselves cool.

There are some native webspinners, a very odd group in which all stages have silk glands in the front feet that gives the forefeet an enlarged, bulbous look. The native species are rare, but one Mediterranean species was introduced early, and really thrives. Young and both sexes of adults may be found in white web tunnels under stones and pieces of wood. All are harmless, feeding on fungi and vegetable detritus. Females are always wingless. Males resemble winged termites to some extent, but lack the ability to break their wings off at the base.

This species is quite harmless and even entertaining if it gets established and puts one of its silken tunnels on a wall. It would find little to eat in most homes.

WEEVILS AND OTHER BEETLES IN FLOUR AND GRAIN

(Family Curculionidae)

Actually, all of the tiny brown insects that invade flour and grain products are not weevils, but they are known familiarly as *weevils* from the old tales of weevilly biscuits on sailing vessels. Adults of all are beetles of one kind or another, brown and slender, less than one quarter inch long. Only those with snouts are weevils. There are

probably larvae in the same product, but these are less conspicuous because they are pale, more closely matching the color of the product. The names of the culprits range from *rice weevil* to *saw-toothed grain beetle* and *confused flour beetle*, to name a few.

Infested products can be freed of the population by heating them in a warm oven until the heat has reached the center of the container. A few hours at 140F will do the job. Long exposure to low temperatures in a freezer can do the same. Exposure to the radiation of a microwave oven seems not to affect them at all unless they are in a product that heats and then the heat kills them. Insects are tough.

Storing susceptible products in tight jars or cans is an effective way of preventing infestation. Once a population has become established, a thorough cleaning of the cupboard becomes necessary. If large numbers of weevils are seen even after heat treatment and cleaning, check to see whether purchased products are already infested. Isolating them in clean cans and jars provides a check as to which products are bringing in the pests.

Stored-grain insects exact a heavy toll on the food supply of the world. Many strategies have been devised to avoid the damage, and for separating the weevils from infested grain. A predatory bug introduced into grain silos in Texas has brought success in controlling these weevils without any pesticide use. Fortunately, our commercial handling has mostly eliminated the problem here. The occasional infested warehouse is an exception. These beetles are universally regarded as pests. One species serves as the intermediate host of a species of human tapeworm in the Orient, but the association has never been reported here.

Insects In Your House

Most houseplants can be grown without pests if you are careful about where they are obtained. If one of your friends gives you a plant or a cutting, place it in quarantine until you are sure that it is not infested.

The easiest way to control many pests is to take the plants outside and spray them with water. If pesticides are resorted to, only those approved for houseplants should be used, and they should be used only out-of-doors. Searching and destroying or eliminating the individual infected plant may be the most practical solution in some cases.

FUNGUS GNATS/SHORE FLIES

(Families Mycetophilidae and Ephydridae)

Tiny, delicate dark flies that hang around potted plants are fungus gnats. These are adults of larvae that live in the soil, feeding on organic matter. The larvae may also invade the tissues of dead plants or plant parts, but are probably not involved in killing the plant themselves, except in the case of germinating seeds.

The main objection that people have to them is that they are present and certainly don't add much to the aesthetics of the potted plant. Cutting down on watering so the soil surface dries between waterings can do some good. Tilling the soil also helps discourage these flies from laying eggs there.

Another small fly, *Scatella stagnalis,* more robust than the fungus gnat, has become a very common nuisance around house plants. Its common name, shore fly, is a bit misleading because it is found mostly in greenhouse settings. The name is used because the family name for most of its relatives, Ephydridae, is comprised mostly of shore-dwelling species. The larvae of this fly feed on blue-green algae that accumulates on the crusty soil of potted plants, and also on roots of some plants. Recent work has shown these flies and their larvae to be likely vectors for the transmission of *Pythium*, a fungus that causes root problems. Turning the soil over periodically helps eliminate the food source for these flies.

MEALYBUGS
(Family Pseudococcidae)

This descriptive name is applied to several different wingless, powder-covered, oval sucking insects that live on various house plants, *Coleus* is a favorite. Some mealybugs have filaments at the tip of the abdomen of the same white wax that makes up their powdery covering. They are very difficult to control chemically once they become established on plants, partly because their waxy covering protects them from contact with pesticides. The species on cacti and stapeliads live primarily on the roots and underground stems.

The life cycle is a simple one, females laying a large number of eggs in a powder-covered mass. These hatch into tiny nymphs that look like the mother. The nymphs move around on the plant until they find a suitable place to insert the mouthparts, and begin to feed. They are able to retract the mouthparts and move again. They differ from scale insects, close relatives, which stay in the same place once they begin feeding. After several molts they reach the adult stage. Adult males have wings and fly to the females for mating, but are seldom seen due to their small size.

The smartest method of control is not to get an infestation in the first place. If you are unlucky, then isolate the first plants seen with the "mealies" as soon as possible. The traditional method of control is searching and killing with a cotton swab dipped in alcohol. The alcohol gets through the waxy powder on the outside of the body. If there is any question about whether bugs still on the plant are alive or dead, they squash juicy if alive. Mealybugs are almost strictly an inside-the-house problem. Be alert because a residual population may be hiding in the soil waiting to launch a second invasion.

SPIDER MITES
(Family Tetranychidae)

Spider mites are common pests of plants grown indoors or in greenhouses. Their activities are insidious, a plant first showing slight webbing, then quickly yellowing and dropping leaves. Spider mites are tiny arachnids that feed by cutting into cells and extracting the contents. As they move around they leave very thin strands of sticky silk, which catch dust.

The presence of dust, an almost universal situation in a house, seems to stimulate their activity. Fortunately, removing the dust often provides a cure. Just take the potted plant outside and give it a good spraying with the hose. If an infestation persists, a pesticide specifically developed for mites may have to be used.

WHITEFLIES
(Family Aleyrodidae)

Whiteflies are tiny white flying insects with sucking mouthparts related to mealybugs and scale insects. They differ from both of those groups in that both sexes of adults have functional wings. The whole body and the wings are covered with a fine coating of powdery wax, which gives them the white color. Under certain conditions in houses, and particularly in greenhouses, huge populations of whiteflies can develop, causing severe yellowing and wilting of the plants. The immature stages have sucking mouthparts, but look like scale insects in being flat and immobile. They may be overlooked on the plant, but are probably causing more damage than the adults.

As in other instances, avoidance of the problem is easier than taking care of it once it has started. Whitefly eggs are very tiny, and inserted into leaf tissue. Because they are so small, it is possible to bring them in with a plant and not notice them. Look for the adults around the plants as an indication of infestation. They fly off the plant readily when disturbed, but do not go far and return to the plant. Chemical control may be necessary to check an infestation, although parasitic wasps can be utilized in greenhouse situations. Yellow sticky traps help ease the adult overpopulation.

Bitten in Your Home

Creatures that feed on us are probably the ones for which we have the least tolerance. It is bad enough to provide a home for an insect, but having it feed on you is too much. More money is spent on controlling the biters than on any other pests except the subterranean termites. Fortunately for us in the dry southwest our houses are very inhospitable to fleas, which are really a nuisance along the coast of California. Brown dog ticks are tough to control but can be gotten rid of in time; they much prefer dog blood to ours. Bed bugs are almost a thing of the past. Kissing bugs are perhaps the most fearsome, and can be dangerous to sensitized people. Mosquitoes are easy to get rid of. Just find where they have been breeding and eliminate the source.

BED BUG

(Cimex lectularius)

Bed bugs have become an oddity these days, but there was a time when people who travelled encountered them in many hotels. This is one of very few insects that prefer human blood to any other. Adults are a quarter-inch long, broadly oval, and as flat as a piece of paper before they have had a meal. Their color is rich chestnut brown before a meal, with a black zone on the abdomen after feeding.

All stages spend most of their time secreted away in some dark crack, inside springs or in the tufts on the mattress on a bed, in nearby wallpaper cracks, in cushions or in the inner recesses of an easy chair. They come out at night if they are hungry, and seek a blood supply from a sleeping person. The wary traveller often slept with the lights on to reduce the chance of being bitten. Infested rooms had a characteristic odor, and there were small brown spots of bedbug excrement near the resting places. The advent of synthetic organic insecticides almost eliminated the problem, but it still exists.

There are a couple of additional species of bed bugs that prefer other kinds of blood, only taking human blood in a pinch. These are smaller than the one that prefers our blood. One lives in the nests of cliff swallows and another in the roosts of bats. Buildings that have either may have populations of swallow or bat bed bugs. Putting a persistent pesticide on nests or roosts while the hosts are away may be necessary if an infestation persists.

BROWN DOG TICK

(Rhipicephalus sanguineus)

People who have moved to the Southwest from other parts of the country may be familiar with wood ticks and tend to extrapolate from them to the situation we have here. Our ticks are almost always brown dog ticks, which are ectoparasites of dogs, and only rarely bite man. Wood ticks are many-hosted ticks. They feed on rodents and other small animals when young and larger mammals including dogs and man when adult. Brown dog ticks are single-hosted, primarily on dogs.

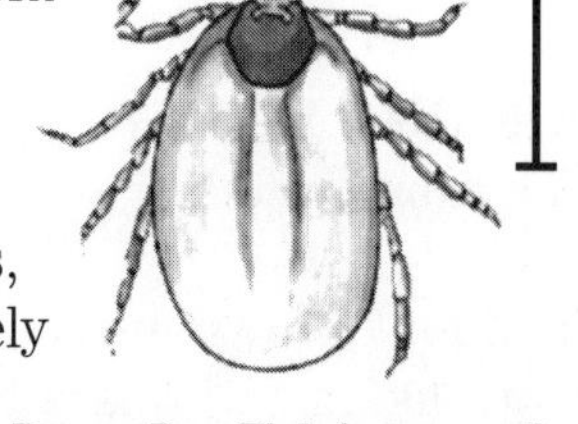

Brown Dog Tick (engorged)
Twice actual size.

Female brown dog ticks lay large masses of eggs in a protected place, often behind a picture if inside a house. The *seed ticks* that hatch from these eggs seek out the family dog for a meal, leaving the dog afterward, digesting the meal, and molting to the next size larger, called a *nymph.* They then go back to the dog for another feed, digest and molt cycle. This third molt results in an adult. Adult females then attach to the host, mate, then engorge themselves with a huge blood meal, so huge that the body is swollen beyond what seems possible. As the eggs in their ovaries mature they change color from the dark purple of the blood meal to brown. The eggs are deposited in one big group, after which she dies. One batch of eggs, if ignored, can result in an unbelievable number of larval or seed ticks, and the cycle starts over. The offspring of one pair can constitute an intolerable infestation.

Because it comes in so many sizes, people have a hard time recognizing this species. The hatchlings are hardly as large as the head of a pin, but get bigger when they feed. Adults start out a little less than a quarter-inch long but the females swell to more than half an inch when they take on a full meal and start maturing their eggs. The general color is rich brown, but the swollen female shows mostly dark purple from the blood in her extensive digestive tract.

Brown dog ticks may be both inside the house and out, so control becomes very difficult. Free-roaming dogs have many opportunities to pick them up. Nothing is really better than not getting the infestation in the first place. All dog owners should at least be alert for the presence of the tiny seed ticks, which are easiest to find behind the dog's ears. Elimination of a heavy infestation may require the help of professionals.

KISSING BUGS

(Triatoma rubida)

Inch-long brown insects that seek out people for a blood meal sound more like a nightmare than a reality. The kissing bug is one of our unfortunate realities. These are insects that have incomplete metamorphosis, the young that hatch from the eggs depending on a blood meal in the same way as the adults. The normal home for the bugs is the nests of rodents, where there is a ready food supply. The familiar pack-rat (white-throated wood rat to the mammalogist) and the armadillo seem to be the usual hosts in the Southwest. Problems develop when the insects become adult and fly around at night seeking mates and a new home. At that time they are attracted to light and may find their way into nearby human habitations. Once there they may not be able to get out, and feed on the nearest mammal, man. Or, they may feed on the family dog. The pack-rat association would keep them away from the most urbanized areas, but urbanization keeps creeping outward to their habitat.

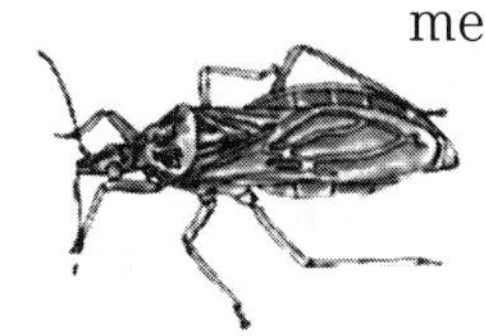

Kissing Bug
Actual size.

They feed only at night, carefully approaching their meal and inserting their mouth parts, usually feeding on exposed skin areas. Because the face is almost always exposed, this was typically a feeding site, thus the name *kissing bug.* The first thing they do is inject an anticoagulant and an anesthetic so they can take on the

meal without disturbing the donor. This injection affects different people in different ways, but most have a hurting/itching welt for a week or more. Some people get sensitized and may go into shock from a single bite. Small pets that are bitten may be so affected that immediate treatment by a veterinary doctor is required.

Good screening and elimination of pack-rat nests in the neighborhood can cut down the chances of being fed on in the home. The usual flight period of the adults starts in May/June, before the summer rainy period, when nighttime temperatures remain in the upper 70s. Placing a bright light well away from the house should reduce the chances of their seeking the dimmer house lights and getting into the house. People camping in pack-rat habitat during this period should provide some protection, even a bed net. Camping in a cave seems to be the surest way of getting bitten.

The kissing bug has come to be known by such interesting common names as the *walapai tiger* and the *conenose* or *coneheaded bug.*

LICE

(Order Phthiraptera)

While lice may not be intimately familiar to many people, they get so much publicity that they are hard to ignore. It is interesting that humans have only three species of ectoparasitic insects that can feed only on us, all *lice.* Many other mammals and birds have more. These insects are so dependent on our blood that they cannot live on any other food.

One of our louse species is very different from the others. This is the crab louse, *Phthirus pubis,* from its very crablike appearance. Adults are about a sixteenth of an inch long and almost as broad, including legs. These live among pubic hairs, and are therefore almost strictly confined to adults, although children may carry them in their eyebrows. These are also known as *crotch crickets* and *crabs.*

Head lice, *Pediculus humanus capitas,* are more slender, no more than half as broad as long, and live in head hair. These have no host age limitation, and can move from person to person very easily. This is the louse that causes most infestations in school children. Both crab lice and head lice glue their eggs to the shafts of hairs,

the eggs hatching and the young lice going from there. All the lice have excellent grasping legs, so are not easily dislodged, even by combing. One of the items of paraphernalia in treating head louse infestations is a nit comb, a very fine-toothed comb that pulls the louse eggs or *nits* from the hair, a variation of nit-picking.

The body louse, *Pediculus humanus humanus,* is closely related to the head louse, and has to have been evolved from head louse stock since mankind took to wearing clothing. These lice spend most of their time in clothing, venturing onto the host only to take a blood meal. They thrive in cold climates and among people living in crowds. During many of our wars they have killed more soldiers and civilians than has gunfire, as they are the vectors of typhus. In peacetime they recede to being the problem of vagrants and refugees, but they are still with us. The discovery of these lice living mostly in clothing led to their colorful nickname, *seam squirrels.* These lice are sometimes called *cooties,* too.

The discovery of the insecticidal activity of DDT ended the threat of typhus by providing a rapid way of controlling lice. A developing typhus epidemic in Naples was halted in less than a week when Allied forces arrived and treated the resident population with a 5% DDT formulation in talc. The standard U.S. military method prior to DDT was steam treatment of clothing while the troops showered.

MOSQUITOES
(Family Culicidae)

This is one group of insects that everybody seems to be able to recognize without prompting. Female mosquitoes of several species seek out the blood of people for food. Most approach quietly and get their meal from small capillaries, on bare skin or through thin cloth. People vary in their reaction, but most have some itching.

The annoying habit of hovering near the ear results from their seeking a place on the skin where the capillaries are close to the surface and the surface is therefore warmer than the surrounding area. Most of the common pest species are night-biters, or start as soon as the sun gets low in the sky. It may be hard to believe, but some species of mosquitoes wouldn't eat our blood if it were the last food on Earth.

Mosquitoes are tied to water where their young develop. Away from home one may encounter biters in recreational areas during most seasons of the year. Night-biters can be overpowering during the summer along the lower Colorado river. Irrigation tail water sometimes generates big populations in agricultural areas. The critical thing is that the water remain long enough for the larvae, called *wrigglers* because of their characteristic movements up and down in the water column, to complete their development.

Lakes with clean shores and a good population of fish may be essentially mosquito-free. Early-season day-biters are a specialty of canyons, at least in the southwestern United States. The common pest mosquitoes lay their eggs on the water surface. Those in the canyons lay them on rocks next to the water. The eggs have to be dried out for a long time before they can hatch. Thus they stay on the rocks over summer and hatch in the early spring when water from winter rain or melting snow reaches them.

Mosquitoes in the home are almost always generated somewhere close to or inside the home. How close can be judged by the presence or absence of male mosquitoes. These look much as the females but have very feathery antennae; they don't fly far from the water where they grew up as larvae and pupae. Males feed only on nectar for their energy source. After a period of rain, water standing in cans and other containers can produce a good crop in a couple of weeks, especially if a few leaves have fallen into the water. Stored rain water serves as well.

Containers filled with well or tap water are usually not to blame, unless they are allowed to get heavily loaded with plant material. The simplest way to take care of the problem is to find the containers and dump them before the larvae become mosquitoes. The familiar larvae transform into pupae, called *tumblers* because of their rolling movements in the water, before the adults emerge. Turning the containers over prevents a new problem from the next rain.

Dog owners should be aware that the common backyard mosquitoes can be vectors of heartworm of dogs, fortunately not a problem for people. One of the species that breeds in larger bodies of water has been implicated in vectoring human encephalitis. When an outbreak of this disease occurs, the need to avoid being bitten goes up dramatically. During the 19th Century the *Anopheles* mosquitoes

Insects That Damage Your House and Its Contents

Fortunately, not many insects pose any threat at all to the home we live in. Subterranean termites are the most important of these. They attack from the ground, and are relatively methodical in their feeding. Dry wood termites and a few species of powder post beetles make up the rest.

In the interior of the house there are a couple of others. Firewood poses the threat of producing adult wood-boring insects that are capable of chewing their way through anything that gets in their way. Then there is the caterpillar of the palm flower moth, which seeks out something to dig into and may select a nice thick carpet for that place. At least one carpet supplier has given up warranting its carpeting as mothproof because of this species. The carpets were inedible to clothes moth caterpillars and dermestid larvae, but the palm flower caterpillar only uses carpet for housing material, not food, so is not affected.

that bred in ponds and marshes vectored malaria to many people. This disease, controlled by the use of DDT treatments for the mosquito vectors after World War II, is once again on the rise in many tropical areas of the world.

LONGHORN BEETLES IN MESQUITE AND PECAN FIREWOOD

(Megacyllene antennata/Neoclytus caprea)

One of the avoidable sources of insect damage in the house is the woodboring beetles that emerge from firewood. Mesquite and pecan are very likely to attract woodboring insects when they are cut, but other woods are also susceptible. The larvae that are boring in the wood usually cut their way near to the surface before making a chamber and pupating. The adults have to be able to cut their way

through the rest of the wood and the bark before they can get out.

The adults, equipped with sharp cutting mandibles, are prepared to get to where they want to go. Most of them are attracted to light, and try to get to windows. The adult beetles are quite noisy when flying and look almost like wasps, causing some consternation to the homeowner, but they are harmless, possessing no stinging device. The obvious way to take care of the problem is to store all firewood outside until it is to be used, especially during the summer.

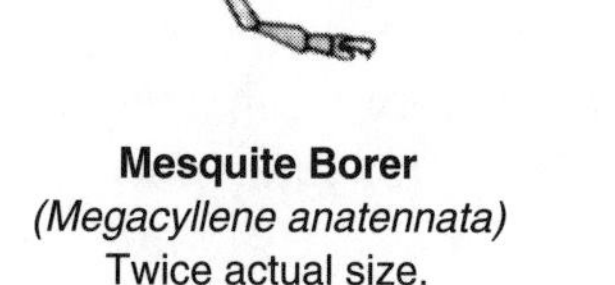

Mesquite Borer
(Megacyllene anatennata)
Twice actual size.

PALM FLOWER MOTH CATERPILLAR

(Litoprosopus coachella)

This caterpillar appears naked, and is usually a striking pink color, with a brown head. It would hardly be noticed except that it may get into a house and make a cavity in a deep carpet or rug. Discovery of a fan palm in bloom inside a patio or at least near the house solves the mystery. The rarely seen parent moth deposits eggs on the palm blossoms as they are opening. The caterpillars feed on the blossoms, ordinarily so far above the ground that they are not noticed. When they have completed their development, they move from the flowers to the bases of dead fronds, stripping fibers to be used in the construction of a cocoon.

A neatly groomed palm will afford no good cocoon material so the larvae wander down the tree searching for a secure hiding place and material with which to build its cocoon. In a paved patio they may not find such a place, and eventually end up inside the house. Because the caterpillar treats the carpet as the upper reaches of a palm, it does not matter whether the fibers are natural or manmade. They are simply cut off and incorporated into the cocoon.

We have had one report that removing the blossoms early in the cycle stopped a recurring problem with this species. Leaving some of the bases of old leaves on the tree also helps by providing a place for the caterpillars to make cocoons close to where they have completed their development.

TERMITES

(Order Isoptera)

Termite Soldier
Four times actual size.

These insects are what many people mistakenly call *white ants.* They are not ants at all, not even closely related. They are small and white, live in large colonies with a queen, and are termed *social insects* like the ants. They produce winged individuals that leave home to find mates and start new colonies. That is where the similarities end. Termite colonies have both males and females in the colony all the time. In fact, the queen 'enjoys' the company of a king most of the time. Termite colonies may be underground or within dry wood above ground in the Southwest. More spectacular colonies are found in Africa and Australia, where mounds range from earthen domes several feet high to major architectural feats 20 or so feet skyward.

Termites are important to the natural world because they have evolved the ability to digest cellulose in plants. With the help of intestinal symbionts, termites act as major recyclers of woody material. Unfortunately, people selected the termites' major food as the material to build homes with. Thus a one-course meal is placed directly above termite colonies in the soil in hopes that they won't just help themselves. In the Southwest, there are many species of termites dependent upon the soil for water and shelter. They venture out in search of nutrition from these sites. Why some houses seem better tasting and subject to termite infestation is an unknown. If one keeps water away from the foundation, and removes old wood from the area, chances improve for keeping the feasters away from your house.

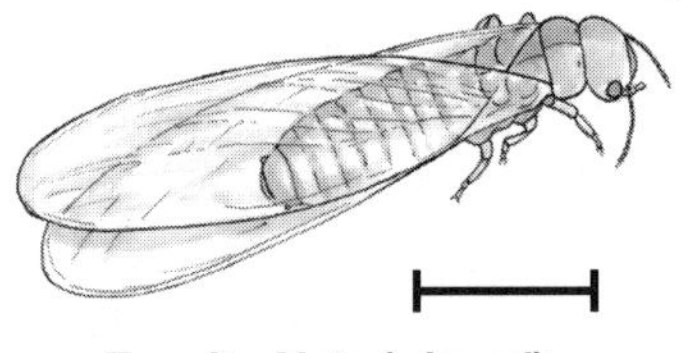

Termite Alate (winged)
Twice actual size.

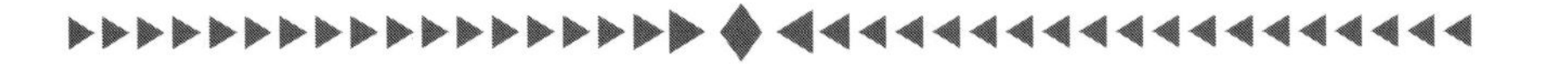

DRY WOOD TERMITES

(Marginitermes hubbardi)

In some other parts of the country dry wood termites are regarded as major pests. The climate of the southwestern deserts is dry enough that they cause little damage here. Colonies are small and contained within wood. They do not have part of the colony underground. The only way that dry wood termites can invade is for a pair of winged termites to burrow into a piece of wood, mate and start a new colony. The start is very slow.

Despite the name, the wood that is invaded almost has to have more moisture than what is in the wood in a house. When wooden floors were popular, the flooring around sinks and toilets was the most likely place for a colony, or wooden window frames that got moisture from rain. A characteristic of dry wood termite burrows is that they are kept very clean. There is no coating of fecal mud. The feces are small pellets which may be stored in abandoned galleries or dumped to the outside through a small hole. Mysterious appearance of piles of fine brown sawdust may be the first indication of an infestation.

In humid climates dry wood termites may invade a house through the attic. Screening of all openings is required in such places. Here the colonies and the damage are more localized and quite noticeable. Removal of the infested wood is the best policy, followed by fixing the source of the leak. New surface and injection treatments of boric-acid compound can supplement the removal of particles in most cases.

SUBTERRANEAN TERMITES

(Heterotermes aureus)

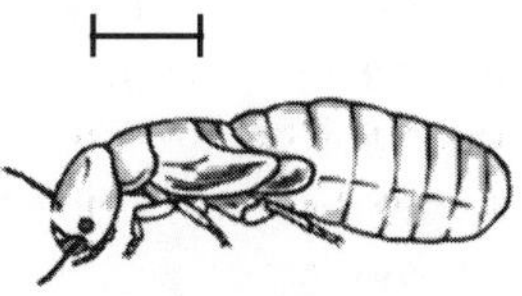

Termite Worker
Four times actual size.

This insect has to be the most ominous threat to the homeowner. Much money is spent on preventing its feeding and on replacing what it has eaten. What the owner of an older home may see first is that wooden baseboards or quarter-rounds cave in when hit or kicked. This is the result of worker termites having eaten the inside parts of the wood, leaving a complete shell on the outside intact. If the feeding is current, slender white insects with darker heads will be seen in the wood. These are the worker termites. The largest of them will be about one quarter inch long.

Subterranean termites line all their burrowings with a gray mixture of fecal material and soil. In newer homes the first sign of an infestation may be in paper or wooden objects stored in closets, or along expansion joints in the garage where mud tubes appear rising along the foundation. Actually, several species are involved in different parts of the country. The main one across the Southwest is *Heterotermes aureus.*

The main colony of subterranean termites is entirely underground. Workers burrow through the soil in search of woody material on which to feed. They concentrate their efforts once a food supply is found. New houses built on an established colony would soon start to become food if there was no barrier between soil and house.

The ordinary practice is to put down a solution of a persistent pesticide before the house is built. If this barrier remains intact, the house is secure. Any breaks in the barrier can be taken care of by soaking the soil with the same material. Periodic inspection is required to make sure that entry has not taken place. In an advanced infestation the workers may construct mud tubes to get over structure that they can't get through, or to explore for new wood. Such tubes may hang from ceilings or connect pictures with walls.

A termite infestation is no cause for immediate panic. Colonies in this part of the world are not as large as the ones you hear about in the wet Tropics. The infestation needs attention, but serious damage does not happen overnight. There is plenty of time to practice prudence in having the building inspected and planning a control strategy. Incidentally, I know of no basis for the statement sometimes made that termites chased from one house are going to look around the neighborhood and select another house as something to eat.

CHAPTER 3 Dwellers in Your Yard and Patio

Once one steps out of the house, the diversity of insects and other arthropods increases dramatically, and the threat to our well-being drops in proportion. With the possible exception of some species that damage food crops, the out-of-door species are very easily left in their natural surroundings, for us to enjoy if we are so inclined.

Among the most colorful are the butterflies. All that must be remembered for these is that almost all of them develop from caterpillars that eat one of the plants somebody is probably growing. Really close attention to pest control on these plants will rid the garden of butterflies.

Likewise, the really diligent pruner will eliminate the tree crickets that enliven our evenings in the summer. Their females lay their eggs in the stems of various shrubs and trees, and the eggs allow the tree crickets to reappear in the spring.

PAVEMENT ANTS

(Forelius pruinosus)

A common sight in all areas is streams of small brown ants crossing the pavement, day after day, all through the warm season. These are the workers of pavement ants. Individually they are very small, about an eighth inch long, and would hardly be noticed if they were not so abundant. Despite the fact that they are following a well-worn smoothed trail, they are really on a scent trail, each ant adding to it as it passes. The experiment-minded person can have some fun with them by interrupting the trail (rubbing a finger across it will do) and observing the confused ants reestablish it. For an even greater disruption a sheet of paper placed on the trail soon has a new one laid across it. The ants can then be diverted by twisting the paper to a new direction. Take it from there.

Somewhere in the vicinity there is a colony headquarters, where the queens and ant larvae are housed (this ant usually has multiple queens). Once in a while one may witness the whole colony moving, workers carrying larvae and pupae, and the queens moving along with them. The queens are more than twice as large as the workers. A characteristic of this group of ants is their odor, something like banana oil or acetate-based glue. Just disturb a few ants and you will smell it. These ants have no stinger, but the secretion gives them a bad flavor. Ripe saguaro fruit lying on the ground is bound to have some pavement ants in it.

This is one of the ants that may invade houses on its foraging expeditions. It is not particularly attracted to sugar or fats, but does general scavenging.

ANTLIONS

(Family Myrmeleontidae)

Antlions
Adult, above (measurement in this example is from head to wing tip). Larva, below. Both shown twice actual size.

This name is given to an insect whose larvae would strike fear into the hearts of everyone if they were as large as cats. They are in the familiar half-inch size range when fully grown, rather rotund of abdomen and bearing a pair of long, curved jaws out front. These are special jaws, because they are hollow. Actually it's not the jaws that are hollow, but grooved mandibles and modified maxillae (secondary chewing mouthparts of insects) lying next to each other in such a way as to make a tube. The insect uses them to inject an anesthetic and digestive enzymes into prey insects, and to suck out the dissolved contents of the body when the enzymes have done their work.

The way in which the antlions catch their prey is what makes

them conspicuous. They seek out fine dry soil or sand and use the head to flip up particles and make a hole. This hole takes on a funnel shape that conforms to the angle of rest of the sand. The antlion larva lies in wait at the bottom of the funnel. When grains of sand land on its head, it flips up more sand. Wandering insects that happen to get into the edge of the funnel have little chance of getting out, especially after that flipped sand and their activities cause a mini-avalanche. When the trapped insect finally slides to the bottom of the funnel, it hits a trigger hair located between the jaws of the antlion, causing them to snap shut, impaling the victim.

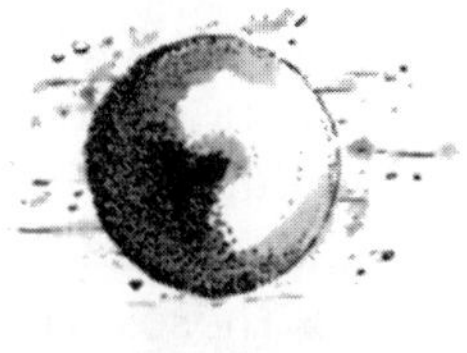
Top View

Side View

The funnels may be as much as two inches in diameter, probably depending on how long the antlion has been there and the texture of the soil. When the larvae move from one place to another, they leave a trail where they have burrowed, usually a tortuous one. New burrows are started as a circular furrow. Excavated larvae move backwards and rapidly begin creation of a new funnel, disappearing within seconds.

Adult antlions seem too large to have developed from the larvae. They are over an inch and a half long, with slender wings held back along the body at rest and short, clubbed antennae. They resemble the more familiar damselflies without the majestic coloration. They are familiar at lights during the summer. They also are predaceous, but are ill-equipped to catch prey. A whole group of species is referred to as antlions, although the ones that dig pits as larvae make up only a part; the rest have larvae that wander around and catch prey in the more ordinary way. The adult illustrated is of one of the pit-digging species.

BEES

(Order Hymenoptera)

Normally the term *bee* brings to mind honey bee, that wonderful insect that produces honey and pollinates flowers. There are many more native bees that make the plant world go round. Here in the

Southwest, bee-plant associations are many. Bees appear incredibly furry, all the better to collect pollen when they go for a meal of nectar deep inside the flower. Next time you inspect that cactus flower, notice what the bees look like that are landing on it.

Be alert when encountering these flying furballs, because the females have stingers that may be used to protect themselves. Most bees act very blustery, and if they fly at you it is bluff. Carpenter bees, those huge black ones, always appear to be on the prowl, looking for trouble, but I don't know anyone who has been stung by them. Watch but don't swat or grab is always a good policy with bees.

BUMBLE BEES

(Bombus sonorus)

Before Europeans introduced the honey bee, the only source of honey in temperate North America was the nests of bumble bees and honey ants. Bumble bees are large bees, the queens being about an inch long, black but so covered with yellow-and-black fur-like hairs that these colors predominate. Some mountain species have some orange fur. These are social bees with annual colonies, which by the end of the season consist of an old queen and as many as several hundred workers. The old colony does not survive the winter in our species.

Mated new queens overwinter in a log or pile of leaves, emerging in the spring to select a pre-existing hole in the ground, often the abandoned nest of a rodent. They make a few cells on the floor of the hole, and use these to hold larvae that hatch from their first eggs. These first larvae are always females, because the queen has allowed the eggs to be fertilized from her store of sperm. She feeds them on pollen and nectar, adding a queen-inhibiting substance from her own body. When they mature they become workers. Workers do the same sorts of things as the queen, using surplus cells for storing excess nectar, modified into honey, and pollen, when there is an excess.

The colony may have several hundred workers at the end of the summer. The last broods consist of females that have not been fed queen-inhibiting substance, and males, the result of the queen

laying unfertilized eggs. Males and new queens leave the nest to spend their time on blossoms and mating. Cold weather wipes out all except the new queens.

Bumble bees are very docile unless the nest is disturbed, but both the queen and the workers are capable of stinging repeatedly, as their stingers lack the barbs of honey bees. They make a colorful addition to the home garden, but are not easily introduced to a new area.

CARPENTER BEES

(Xylocopa californica)

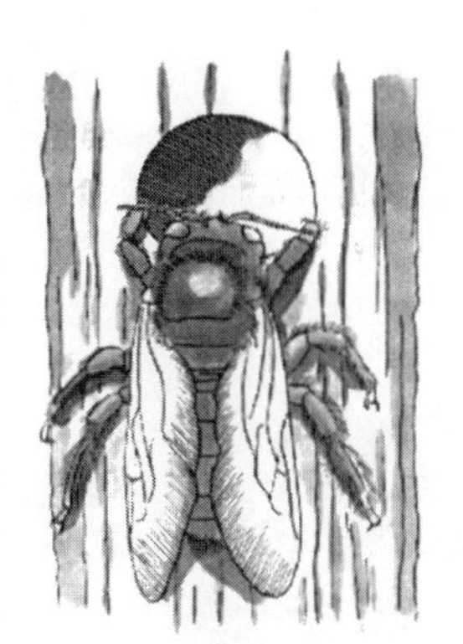

Carpenter Bee
Actual size.

Carpenter bees are the only really large bees in the Southwest that are metallic blue-black to black (one species has a yellow male). The commonest ones are about an inch long.

Carpenter bees are counted as solitary bees. This means that each female makes her own nest and provisions it for her own young alone. We have several species. The commonest one from Texas to southern California, and the one figured, prefers the dead flowering stalks of sotol or desert spoon as a place to put its nest. The female burrows into the stalk for some distance, makes cells, provisions them with a paste of pollen and modified nectar, lays an egg, and goes on to make another. Two other species prefer to make their burrows in solid wood, especially chinaberry trees. One of these is a large species, *Xylocopa varipuncta,* with males that are yellow rather than black. The other is smaller, with both sexes black. The smaller one ranges north into at least the juniper zone.

In the Tropics there are solid-wood species that are so diligent that they can affect the stability of a structure. Because our common species uses dead sotol flower stalks, our carpenter bees are quite harmless. For those not fortunate enough to have flower stalks in the yard, the bees are easily encouraged by cutting a sotol stalk or two and leaning them against a wall. If there are any carpenter bees in the neighborhood, at least one female will start burrowing into a stalk, always on the underside. She buzzes while she works, and will soon have a pile of white sawdust accumulated.

Almost certainly at least one carpenter bee will establish territory around the nest, chasing other bees, birds and even people. This will be a male carpenter bee, harmless because he lacks a stinger. If there are no bees in the neighborhood, they can be moved in by cutting a freshly burrowed stalk during the winter, and bringing it in. Emerging bees will turn up at about the time that Palo Verde starts to bloom.

A very useful invention of hunters in the Southwest has been the *Moses stick*. This is a straight sotol stalk cut exactly the right length to use as a rest for binoculars in glassing a hillside. It is smoothed and the most deluxe models are fitted with a crutch tip on the bottom end. In our family there are usually several that get stuck in a corner of the yard when not in use. Carpenter bees find them very attractive. Now we regard no Moses stick as official unless it has been used once by a carpenter bee. The bee is a little less welcome when it decides that a leaning broom is a sotol stalk. We have never had one make a nest, but several have started and gotten discouraged.

LEAFCUTTER BEES

(Megachile sidalceae)

The mysterious disappearance of circles and ovals from the edges of leaves is the work of female leafcutter bees. These are solitary bees, each female making and provisioning cells for her larvae to develop in, as does the carpenter bee. Leafcutter bees are of honeybee size or smaller, black or gray with white hair forming bands on the abdomen, and with stiff yellow bristles forming a pollen brush on the underside of the abdomen in the female. The male lacks the brush and has front feet like white boxing gloves.

Leafcutter Bee Work

When the female is ready to construct a cell, she searches for a hole of appropriate size, slightly more than the diameter of her body. Then she constructs a thimble-like cell of oval pieces of leaf or petal, usually using at least half a dozen pieces. Then she heads out to gather pollen and nectar from blossoms, to make *bee bread* to provision the cell. Finally she lays an egg on the bee bread and proceeds to close the cell with circular pieces of leaf. The whole job can be done remarkably rapidly, usually in the middle of the day. After finishing a cell she is likely to take care of her own needs and spend some time in the nest entrance looking out. One cell per day seems to be normal.

Roses, bougainvillea, ash, redbud and many other plants with thin, smooth leaves or petals are selected. Protecting the plants with pesticides is difficult, but apparently is possible with persistence. Eliminating nest-size holes would be more permanent. Large nail holes would be the main ones, but nests may even turn up in rolled-up blinds. If annoying the insects is one of your pastimes, try leaning a piece of pegboard against a wall and count the number of times a female leafcutter has to pass through before she decides that the holes are not appropriate.

One small leafcutter bee, accidentally introduced into North America from Europe, is raised for use in alfalfa pollination in several states north of Arizona. The females of that species are willing to nest close to each other. The first design of nest boxes consisted of boxes of milkshake straws in a shelter. The usual design now used depends on numerous holes drilled into a timber. The filled cells can be held in cold storage until a new generation of pollinators is needed. Leafcutter bees are much more efficient at pollinating alfalfa than worker honey bees, which learn after several visits that the alfalfa blossom hits them on the underside of the head with the anthers.

If you are not really concerned about the cutting done by these bees, the females can often be induced to make their nests in holes drilled into the surfaces of two boards screwed together. The contents of the nest are easily examined by pulling the boards apart. Then such questions can be addressed as why the adults in the deepest cells, which are the oldest, do not destroy their siblings as they make their way out of the nest. You may also discover that some or all of the cells are occupied by predators or parasites. Science fair project? Start early in the summer, because it may take some time to find the right hole diameter. One particular species of leafcutter bee, *Lithurge apicalis,* has become expert at using foam roofs as nesting sites. A small imperfection can quickly be expanded for a nest and filled with numerous nursery cells. This will in turn attract the flickers and soon your nice white roof will be pock-marked. Yearly inspection and maintenance are a necessity to prevent this damage.

BUTTERFLIES
(Order Lepidoptera)

These are the jewels of the insect world. Everyone seems to love butterflies because of the radiant color pattern of scales on their wings. These showy animals belie what they once were: fat, juicy caterpillars. Butterflies are brightly colored because they are diurnal creatures and their colors can be seen by all. Butterflies elicit no ill feelings by people because they don't bite, sting, vector diseases or become nuisances around your face.

Everyone enjoys the summer show of colors that these wonders of life exhibit.

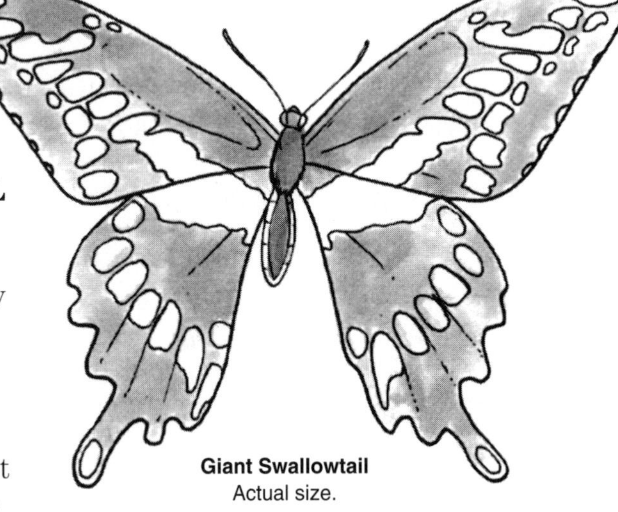

Giant Swallowtail
Actual size.

GIANT SWALLOWTAIL
(Papilio cresphontes)

By far the largest butterfly in the urban areas of the Southwest is the giant swallowtail. With a wingspan of about four inches, it is the largest insect most people will ever see in the wild. We do have some larger moths in our forested areas, but these night-flyers become active after midnight. The color pattern on the Swallowtail's wings is distinctive, yellow-brown with a yellow band

on the upperside, mostly yellow on the underside. Many related species are widespread in the Tropics, and we have several swallowtails native to the Southwest.

The giant swallowtail is also a good example of the benefits of permitting a few insects to feed on a crop. The caterpillar feeds exclusively on citrus, consuming the younger leaves with great gusto. After this feast and time to change clothes, this beautiful butterfly provides joy and excitement to people in their backyard. When a giant swallowtail is cruising around your orange tree, it may be a female looking for places to deposit some eggs.

The caterpillar is as distinctive as the butterfly. It looks exactly like a fresh bird dropping, complete with the white markings of the uric acid that birds excrete, and even shiny. If it is poked, it shoots out an orange "forked tongue" from behind the head; the tongue emits a rather sweet-smelling defensive secretion. The tongue and smell are characteristic of all swallowtail caterpillars.

Except in the case of very small trees it is unlikely that enough caterpillars will survive on a citrus tree to cause measurable damage. A complex of parasitic insects seems always to be present, killing most of the eggs and young caterpillars. A very fussy gardener might object to part of the leaves having pieces eaten out of them. There have been places where the caterpillars, called *orange dogs,* have done significant damage to citrus, but these have probably always been where other pests such as scale insects have been treated with pesticides.

GULF FRITILLARY

(Dione vanillae)

Gulf fritillaries are large rich-orange butterflies with black markings on the upperside, plus conspicuous silver spots on the underside of the wings. They can be fairly abundant in our major urban areas, but are rarely seen in the desert. The simple reason is that the only food plant that is acceptable to the caterpillars is the passion-flower, several species of which are grown as ornamental vines. The group to which this species belongs is a big one in the American Tropics, but only the gulf fritillary is found commonly across the Southwest. It almost certainly was introduced with passion-flower plants.

PIPEVINE SWALLOWTAIL

(Battus philenor)

Pipevine swallowtails are not as large as giant swallowtails, being about three inches in wingspread, but with the same characteristic wing shape. The upperside of the wings is mostly a rich, iridescent blue; the underside is blue with orange spots. This color pattern makes the butterfly conspicuous on almost any background. The predominant color in most swallowtails is yellow, one that matches the background when they are on blossoms and even among leaves.

Studies of pipevine swallowtails and their relatives in the American Tropics have shown that there is a simple explanation for the color pattern. These butterflies are highly poisonous and distasteful, even to birds. The distinctive color identifies the butterfly as something not fit to eat for any bird that has tried one before and learned the hard way. The fancy term for this is *aposematic* or *warning coloration.*

The poison in the body has come from the food of the caterpillar, the Pipevine or Dutchman's pipe. A similar situation occurs in the case of the more widely publicized monarch butterfly, which eats milkweeds in the larval stage, sequestering the toxin in its body and carrying protection on to the adult stage. Pipevines are sometimes planted as ornamentals, but not widely. We do have a native desert species that sprouts from a tuber every year. The female pipevine swallowtail has such a good sense of smell for pipevines that one growing without swallowtail caterpillars is a rarity. Most often one sees the caterpillars, which are dark purple with orange filaments or sometimes entirely orange, and possess the hidden "forked tongue" scent sprayer of all swallowtails. There are no pipevine plants visible. They have all been eaten and the caterpillars are probably out searching for more or possibly a suitable place to become a *chrysalis,* the pupal stage of butterflies.

CICADAS

(Family Cicadidae)

Cicadas are sucking insects with incomplete metamorphosis. They are large, from one and one-half to two inches long as adults in the several species that are abundant at lower elevations, to more than

two inches in some species found in the oak zone. The commonest species at lower elevations is the Apache cicada, which has a body that is mostly black (to tan in the Yuma region), with a pale band just behind the head.

Cicadas are conspicuous because of the males' mating calls during a summer's day. Apache cicadas emit a continuous high-pitched call that some people find annoying. Others have a call that pulses. Each species has its own call that serves to attract the female of the species.

Cicadas
Cicada (Dog Days) above; cicada nymph below, both actual size.

Cicada nymphs spend their entire lives underground, making smooth tunnels that they use to get to the roots of plants, into which they insert their sucking mouth parts. They spend several years underground, molting and growing to nearly adult size. In the year when they will become adult, they burrow up near the surface and spend some weeks there before emerging from the ground. At the last stage the nymph, which has a fully formed but soft adult inside its exoskeleton, emerges from the soil and climbs onto something for the last shedding of the exoskeleton. This emergence occurs shortly after sunset and the adult will have emerged, pumped up its wings and hardened enough to fly by midnight. By dawn it is ready to go. This is a good thing for the cicada because it is an eminently edible object. Some are probably eaten by nocturnal predators before dawn, but daylight brings the eager attention of birds and lizards. Even seed-eating birds have a cicada feast.

Watching the emergence of the nymphs and the expansion and hardening of the adults makes an interesting diversion on a warm evening, and could serve to alleviate the fear of the dark that seems to be part of some children's lives. The soft nymphs make very good food for insect-eating pets, and can be stored in the refrigerator or freezer.

Adults also feed on plant sap with their sucking mouthparts. Some species are quite particular about the plant, but the common

urban species seem to use about any tree or shrub. After the singing has resulted in mating, the females insert their eggs into slender twigs in a row. The physical damage to the twig is often severe enough that it dies at that point. Usually this natural pruning is not detrimental to the plant, an exception being a young seedling that may become stunted. The hatchling nymphs drop to the ground, dig down, and restart the cycle.

The Apache cicada, *Diceroprocta apache,* seems to have a three-year cycle. This species staggers its adult emergence, so it appears each year, leaving behind developing nymphs underground. Those nymphs will emerge in the following two years. This adaptive behavior is one of many ways insects ensure survival of the species. If environmental conditions or some catastrophic event occurs to eliminate the emerging population, the individuals still developing in the ground will survive and emerge the following years and thus continue the success story.

In the northeastern and midwestern U.S., the famous 17-year cicadas, *Magicicada* spp., (mistakenly called *locusts,* a term used to describe specialized grasshoppers) have enormous broods on a 17-year cycle. This behavior, putting all your eggs in one basket, takes the opposite approach from the Apache cicada, but is still equally successful. In the Southeast very similar species emerge on a 13-year cycle, possibly the result of warmer climate. Even more amazingly, there are three extremely similar species involved in both the 17-year and the 13-year cicadas, differing in the calls of the males and the preferred habitats.

DESERT COCKROACH
(Arenivaga genitalis)

An insect that is recognizable by its shape as a cockroach but not by its habits is the desert cockroach. This is a flying insect about three quaters of an inch long, with a delicate pattern of brown on a tan background. The individuals that are seen happen all to be males, that are nocturnal and very strongly attracted to lights. The females are completely wingless, more broadly oval and much shorter, perhaps half an inch. These sometimes turn up in a yard. They probably are attracted to lights also, but don't get to them because they have to walk.

Desert cockroaches spend most of their lives burrowing in soft soil, probably feeding on organic debris. They are fairly abundant in pack-rat nests, where the soil has been loosened and there certainly is a higher concentration of organic matter than elsewhere in the desert.

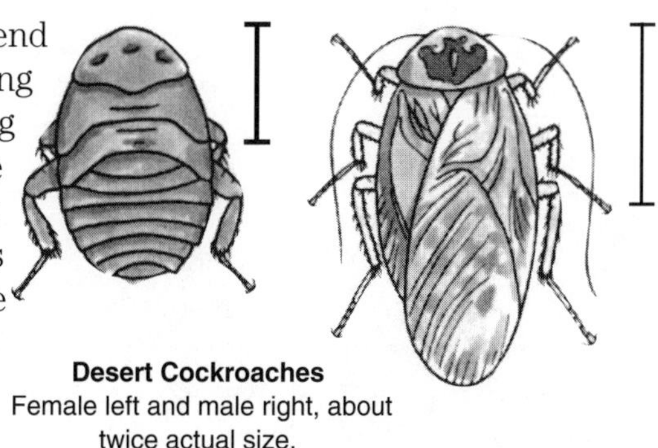

Desert Cockroaches
Female left and male right, about twice actual size.

Desert cockroaches are harmless, of course, and would not even be very likely to get inside a house unless strongly enough attracted to fly in when a screen door was opened.

CRICKETS

(Order Orthoptera)

Crickets are relatives of the familiar grasshoppers. Crickets, though, have evolved a communication system which most everyone has heard at one time or another. Crickets do not rub their legs together to talk. Rather they use wings rubbing highly adapted parts together to produce those nightly songs. Generally, people hear the males singing in hopes of attracting a mate. Sit outside on a summer's evening and listen to the chorus of song produced.

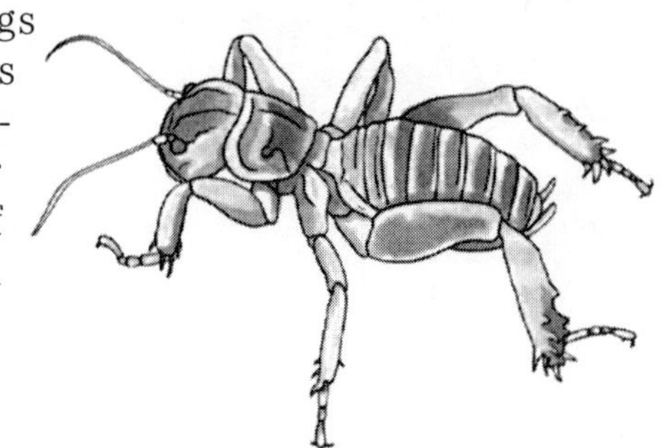

Jerusalem Cricket
Actual size.

JERUSALEM CRICKETS

(Stenopelmatus spp.)

The Jerusalem cricket is a large, tan, cricket-like wingless insect with an oversized head. The head is so oversized, especially in the male, that it looks much like a tiny human skull; hence the local name of *child-of-the-Earth,* or its Spanish equivalent, *Niña de la*

Tierra. A certain amount of superstition has developed around it.

The insect is strictly nocturnal, and can be fairly abundant in an area without being seen. Fully grown ones are an inch and a half long. They burrow into the soil, where only a gardener would be likely to encounter them. Most of the ones that are picked up had wandered into swimming pools.

All stages are actively predaceous, consuming whatever they can catch in their size range. The large head accommodates large muscles to work the jaws, so they can deliver a meaningful bite if one picks them up. They do well as pets, although their nocturnal habits keep them out of sight most of the time.

TREE CRICKETS

(Oecanthus spp.)

Tree Cricket
Twice actual size.

Tree crickets are slender green crickets about three quarters of an inch long that spend the day on green plants, where they are extremely well camouflaged. At night they are usually the loudest of the noise makers, one species having a melodious chirp-chirp-chirp-chirp-chirp song that goes on for ages, and another a continuous call that goes on just as long. The calls are made by the males. They raise up their front wings to do it, scraping a ridge on one wing across a file-like device on the other wing. The field cricket and the Indian house cricket sing in the same way.

These insects are leaf-eaters of the shrubs that they are on, and the females have slender ovipositors, which they use to insert their eggs into twigs, in much the same way as female cicadas, but rarely to the detriment of the plant. The usual life cycle is an annual one, with the winter being spent in the egg stage.

Big populations of tree crickets are usually an indication of rampant shrub growth in a neighborhood, the shrubs going from year to year without severe pruning. Most people find the cricket calls a

welcome addition to the noises of the city. Incidentally, the chirpers respond to temperature and one can approximate the temperature in degrees Fahrenheit by adding 40 to the number of chirps in 15 seconds. Several chirpers, known as *snowy tree crickets,* may call in unison for fairly long periods.

DADDY LONGLEGS

(Order Opiliones)

This group of arthropods belongs to the class Arachnida, close relatives of spiders, scorpions and ticks. All have an oval body up to a quarter inch long and extremely long, slender legs. The legs readily break off if caught in a spider web, a slick escape mechanism rivaling the lizard's lost-tail trick. Their arachnidism is indicated by the fact that they have four pairs of legs and lack the antennae that are characteristic of insects.

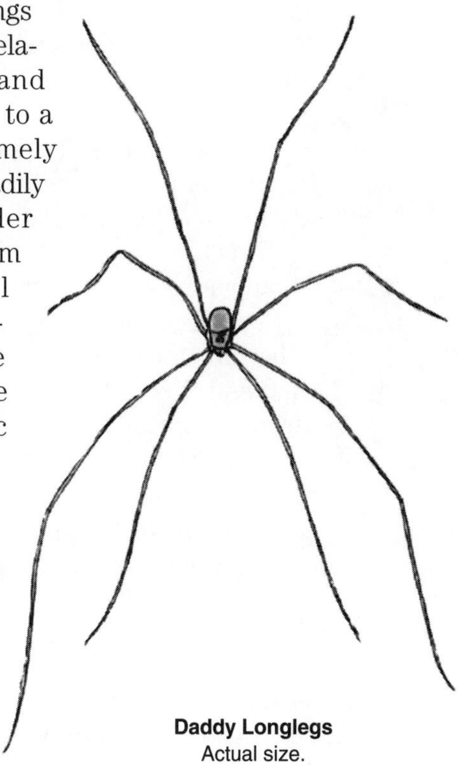

Their mouthparts are of a chewing type similar to scorpions, not fang-like as in spiders. Daddy longlegs are characteristic of damp places and cavities. They are an indication that a garden has reached a certain stage of maturity.

Daddy Longlegs
Actual size.

Their main role is as predators on insects and other small arthropods. All protect themselves with an evil-smelling secretion. Unlike spiders and scorpions, they do not have a venomous bite or sting. Females lay eggs in small masses, and the young look like diminutive adults.

The scientific name for this order of arachnids, Opiliones, is derived from the word Opilio that means "a shepherd" in Latin, likening the long legs to the stilts shepherds used to walk about on when counting their flocks. In England these animals are known as *Shepherd spiders* or *harvestmen,* and the French call them *reapers.*

EARWIGS

(Labidura riparia)

Earwigs are slender, brown insects with a pair of pinchers in the rear. We have several kinds, including the one figured, that is one and one sixteenth inch long with paler markings. These are chewing insects, distantly related to grasshoppers, and have incomplete metamorphosis. Baby earwigs look like small adults, but the youngest ones have straight, slender appendages at the tip of the body instead of pinchers. Several of our species are winged in the adult stage, but the functional hind wings can be neatly folded up like a fan and then pleated so they fit under the much shorter front wings, leaving only a small portion exposed.

Earwig
About twice actual size.

Earwigs live in damp places and are nocturnal. Males differ from females in having the pinchers at the tip of the abdomen well separated at the base, variously toothed along the inner edge and usually much bigger. They use the forceps to hold the female during mating. When they are ready to lay eggs, females construct a chamber under a stone and lay a mass of shiny white eggs. They practice maternal care of the eggs, circling the body around the egg mass until it hatches. The young soon leave the chamber. The food of most species is dead plant material. But the one figured is an active predator, catching prey with its pinchers and consuming it with its rather small and feeble mouthparts. All species are provided with a bad-smelling secretion from the sides of the abdomen. The name is another that came from England, and derives from an ancient superstition that earwigs invade the ears of sleeping people and drive them insane.

DESERT GRASSHOPPERS

(Trimerotropis spp.)

Desert grasshoppers are often the most conspicuous insects that accumulate in lighted parking areas in the lower desert. In shades of tan that match the desert soil, and with yellow and black on the hind wings when they fly, they are very distinctive grasshoppers. They are about an inch and a half long.

They belong to a group of grasshoppers called *band-winged,* because all of them have some kind of markings on the hind wings, obvious when they fly and completely hidden when they land. The bright colors probably serve two important functions. One is called a *reverse-flash protection.* What a bird sees when they are in flight is the brilliantly colored hind wings. If the grasshopper manages to elude the bird in flight and lands, what the bird was chasing suddenly disappears. The confused bird flies on and the insect is safe.

The other function is *mate recognition.* There may be several different band-winged grasshopper species flying at one place. The particular wing pattern, reinforced by some differences in behavior, enables the females to recognize their own show-off males. The males of some species clack their wings on their legs as they fly to call further attention to themselves. One that is adult in the fall sounds startlingly like a rattlesnake.

Most grasshoppers hatch from eggs in the spring, are in the nymphal stages all summer and mature in the fall. The first freezing weather kills all but the eggs laid for the next year. A few of these have more than one generation per year. Band-winged grasshoppers can overwinter as nymphs, and even as adults. The desert grasshopper is adult in early summer. There have been instances in which populations have damaged seedling crops. But adults that fly to light at night don't seem to cause trouble to adjacent gardens in the morning.

An interesting facet of their behavior is their orientation in relation to the sun. On cool mornings they can be seen sitting perpendicular to the sun, obviously catching as many rays as they can. By midday their orientation is parallel to the sun, so that they get minimum heating. The effect is very noticeable when there are hundreds of grasshoppers on one wall. The local birds usually find the grasshopper bonanza and have a big feed. The grasshoppers are so strongly phototactic that their numbers are replenished the next night.

MOTHS AT LIGHTS

(Order Lepidoptera)

Many different nocturnal insects are involved in the swarms of insects that fly to lights and congregate around them. Collectively they are often referred to as *moths* or *millers*, the latter term probably tracing back to the flour covering the grinders of wheat looking like the pale colors on the moths. The influx of insects is greatest on warm, humid evenings, with the timing often correlated with the onset of the summer rains. The flying ants and termites that are also involved are discussed separately.

Certain lights bring in more insects than others. Fluorescent tubes are the most attractive. This has been demonstrated to be the result of these tubes producing light in near-ultraviolet bands. Insects happen to see far into what we call the *ultraviolet*, a term really reflecting our human bias (beyond what we can see). Very few of them see much at the red end of the spectrum. An insect naming the colors would probably refer to what we see as red in terms of near infra-red. So fluorescent tubes seem very bright to insects. The yellow lights sold as insect repellents are really just not strongly insect attractant. Red lights would be even less attractant, but we see red rather poorly ourselves and the red light has certain unfortunate connotations.

Why an insect that is flying around at night, looking for a mate, a place to lay eggs, or a bite to eat, should give up the search and come to the light is something else. What seems to be the most logical explanation is that they tend to use any point of light as a basis for orientation as they fly. This would be the moon, a planet or a bright star in a natural habitat. By keeping the light at a particular angle to the eyes, they would be flying straight. However, if the light source is much closer than the moon, keeping the same angle causes them to spiral in to the light, getting closer and closer, finally hitting it. Before the invention of electric lights, they did the same thing to open flames. In fact they still do.

Most nocturnal insects have eyes that adapt to light intensity by the movement of pigment. They can detect tiny amounts of light when dark-adapted, but shut out most light when there is a lot. To them bright light means daytime, when most good moths go to sleep. So they spiral in to the light, get their eyes light-adapted, go to sleep and are trapped there.

Many moths have another specialization of note. When their eyes are dark-adapted, they glow when a light is shined directly at them. The pigment movement exposes a mirror layer deep down that makes it possible to run the rays through the receptors on the way in and the way out. "Shining" eyes is an interesting nocturnal pastime. Just hold a flashlight near your eyes, so the light goes straight to the reflector and straight back to your eyes. Not only moths but some other insects and many spiders have reflective eyes, as do most mammals and some birds. Even a few people, whose eyes glow pink when the flash is held in the wrong place in a photography session, exhibit this trait.

PRAYING MANTIDS

(Stagmomantis spp.)

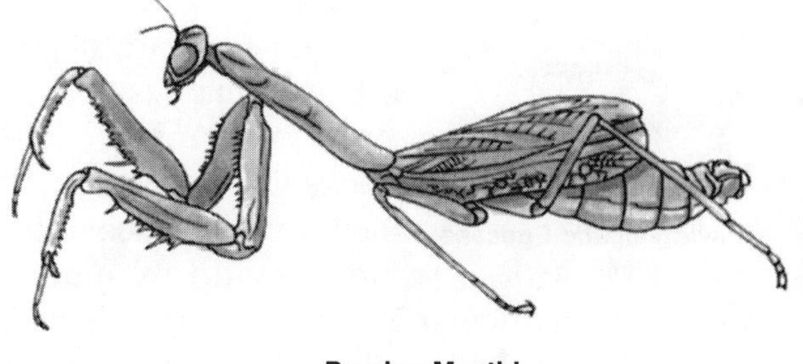

Praying Mantid
Actual size.

Just about the most famous predatory insects are the praying mantids. Almost as famous is the "knowledge" that they are protected by law, which they seem to have been once for a short time in New Jersey, when an exotic species was being introduced. They are not now, as far as I know.

The Chinese mantids, *Tenodera aridifolia sinensis,* familiar to people from the eastern United States, are green and two and a half to three inches long as adults. The Southwest has a wider variety than that. Several desert species are smaller, narrower, and brown like the desert soil. We do have some large green ones, varying to yellowish or tan. In all the Southwest has about ten native species and one that has been introduced from the Mediterranean. This last one is spectacular in the adult stage because the hind wings have a large dark eye spot. When the mantid is disturbed, it raises its wings straight up and rubs its abdomen on the wing veins. Literally translated, its scientific name, *Iris oratoria,* means "talking eye," an apt description. Another unique species is the unicorn mantid,

Pseudovates arizonae, so named because of the single horn emanating from between its large eyes.

Mantids grab their prey with special raptorial front legs that can be shot out rapidly to grasp another insect. Their mouth parts are grasshopper-like, not at all suitable for a predator. Even the mandibles are small. The mantid must hold its prey firmly while it eats it like a stalk of celery. The mantid has no anesthetic or venom to inject, eating its prey alive. To give the reader an idea of a mantid's power, speed and strength, several cases are known of mantids catching and eating hummingbirds. Mantids wait in ambush, counting on their concealment, excellent vision and rapid reflexes to make them a living. The head is almost on a swivel, and can be turned almost straight back. Courtship and mating can be a hazardous undertaking for the male mantid. The female is likely to grab him and start eating from the head end.

Praying Mantid Eggcase
Actual size.

Mantid eggs are deposited in characteristic oothecae, masses of eggs covered with hardened foam. Young mantids emerge from what looks like a zipper along the top. If there are holes on the sides of the ootheca, these have been made by emerging small parasitic wasps. The hardened froth of the egg case keeps most parasites out, but one wasp has evolved the capability of inserting its eggs. Amazingly, there does not seem to be much indication that birds have learned about the edible contents of the conspicuous egg cases.

Egg cases brought into the classroom in early spring can result in hatching of young mantids by the dozens. These must be provided with such food as laboratory fruit flies or there soon will be only one slightly larger young mantid in the jar, perhaps introducing the reality of the predator-prey relationship at an inconvenient time.

PUNCTURE VINE WEEVILS

(Microlarinus lypriformis) (stemborer)
(Microlarinus lareynii) (seedborer)

The insects discussed here are not conspicuous at all, and what they do is also inconspicuous. Two small weevil species were introduced from Cyprus in the Mediterranean to serve as biological

control agents of the puncture vine, *goathead* or *bullhead* as it is known to old-time Arizonans. This is a prostrate weed that at one time was very abundant at lower elevations throughout the Southwest. The plant is quite attractive, bright green and forming a mat over many square feet in a wet season. The small yellow blossoms are also attractive, but the fruit that soon develops is a terror. It is made up of four parts, each with two long, very sharp points. The spines are strong enough to penetrate deeply into a bare foot, and through most bicycle tires.

The plant was introduced from the Mediterranean many years ago, so the Mediterranean was a logical place to look for insects specialized to feed on it, a standard procedure in biological control. First the weevils were tried out on our important cultivated plants and found not to affect them. They were released here, with people picking up infested plants to move around once they got started. In a few years they really reduced the puncture vine. The weevils and their larvae look alike, but the larva of one attacks the plant by getting into the crown and eating it out; the other gets into developing seeds.

An unexpected impact that infuriated some botanists was that the species that gets into the crowns does the same thing to the related Arizona poppy, *Kallstroemia grandiflora.* This native plant, once an abundant roadside attraction during the summer, has become much scarcer. At this point it is hard to say whether the parents who purchased thorn-proof inner tubes for their childrens' bicycles would have prevailed over the botanists if this impact on the Arizona poppy had been known in advance. The matter is now moot. The puncture vine is still with us, but scarcer. The Arizona poppy is also, but not quite as common as it once was.

SOWBUG/PILLBUG

(Porcellio laevis)/(Armadillidium vulgare)

The creatures covered by these names are arthropods but not insects. Instead they are classified as *crustaceans,* along with crabs and shrimp. Almost all crustaceans live in the ocean, a few in fresh water, and a very few on land. Sowbugs and pillbugs are about three eighths of an inch long when fully grown. They can live in the desert, but betray their crustacean ancestry by

Sowbug
Twice actual size.

being quite bound to sources of moisture. One finds them under pieces of wood and stones, from which they forage at night for organic matter. Pillbugs are so designed that they can roll up into a ball when disturbed. Sowbugs lack this ability. Otherwise they are very much the same. Most of the widespread species which are found around homes in the U.S. are of European ancestry.

Pillbugs have been implicated in damage to seedlings that have just emerged from the ground, but otherwise these animals have to be regarded as beneficial in breaking down dead plant material and as a continuing supply of animal matter for the birds and lizards that search among dead leaves for their food supply. They make quiet pets and their avoidance of light can serve as the stimulation that makes them good racing animals on a dull afternoon.

ENCRUSTING TERMITES

(Gnathamitermes perplexus)

Because subterranean termites can be a serious threat to the homeowner, it is easy to become concerned about other termites in the yard. Even a close examination of the workers and the much more distinctive soldiers might indicate that the damaging subterranean termites and encrusting termites are the same to the uninitiated. The prothorax, which forms a rather broad collar behind the head, is slightly saddle-like in encrusting, flat in the subterranean termites. To make matters more confusing, encrusting termites also have the main part of their colonies underground.

The main point of difference is that encrusting termites have a much reduced capacity to eat into solid wood. The workers are able to scrape off pieces of highly weathered woody material, or such things as dry grass. They are unable to handle solid wood. The way in which these termites operate is to come to the surface and seek out soft materials. If they find a weathered stick, they cover the entire surface with a fecal mud chamber and consume the weathered surface before moving on to other sticks. When they eat a blade of dead grass, they may leave a slender tube standing upright in the desert.

On a palm trunk or a saguaro they may mud plaster several square feet of surface at the bottom. All they are doing is cleaning up the old papery layer of dead wood. They do not harm them. These insects are abundant all across the warmer parts of the southwestern

desert areas. Their surface activity is most obvious following rain during the warmer months, but may continue in the winter on warm days. A good spring growth of annuals leaves much for these termites to consume in the fall.

WASPS

(Order Hymenoptera)

Wasps are the naked relatives of the bees. Wasps may also bring thoughts of sleek-waisted individuals flying to some kind of paper-nest site. These insects have much less hair than bees, because they are more hunter-oriented than flower-oriented. Most wasps, females again, have a stinger that is used to paralyze other insects which become food for either adult or juvenile wasps. People encounter the stinger when they come too close to mama's offspring (the nest) or if one panics at the sight of a wasp and starts swinging. Wouldn't you try to protect yourself if a giant tried to hit you?

Wasps build all types of nests, from open or enclosed paper nests to underground cells or mud cells which they fill with paralyzed spiders or insects. Here again one finds that wasps help counter overpopulation of many species of insects one might not want around the garden.

PAPER WASPS

(Polistes spp.)

Paper wasps are like bumble bees in that they are social insects that live in a colony of related insects, and the colony is also annual. The wasps themselves are familiar because they are seen on the paper nests that they construct. Most are about an inch long and fairly slender, with a body in shades of brown to black, and yellow.

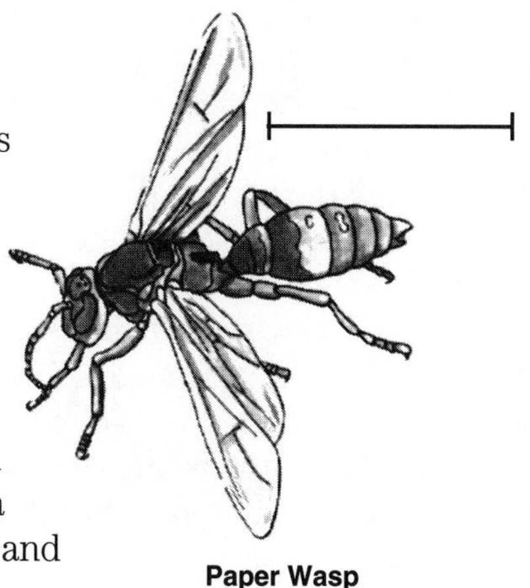

Paper Wasp
Twice actual size.

Colonies are started in the spring by overwintered mated females, as is the case with bumble bees. The start of the nest is the selection of a site in a sheltered place under a roof, under dead palm fronds or inside a bush like pyracantha. Here the female makes a cell of paper, suspending it from a short pillar. Unlike the bumble bee, several overwintering females are likely to start cells together, each making her own cell. When the cells are completed, the females lay eggs in them. It appears that the ordinary thing to happen is for one of the females to become the de facto queen of the colony by systematically eating the eggs of the other founding females and substituting her own. When the eggs hatch, the larvae are fed chewed-up insects and spiders. Because the cells are vertical, it is critical that the larvae are sticky. They don't fall out. When they are ready to pupate, they spin silk caps to close off the cell.

The queen lays fertilized eggs that become females. The larvae are fed queen-inhibiting substance, which suppresses the development of their reproductive systems. The emerging workers look exactly like the fertile, founding females. They make cells and feed larvae just like the founding females.

As in bumble bees, fertile females and haploid males (from unfertilized eggs) are produced at the end of the season. These leave the nest and mate, the new fertilized females being the only individuals that live through the winter.

Nests constructed over a screen door on a porch, or in shrubbery that needs pruning, can become a nuisance. The wasps are highly diurnal creatures. If one decides to try to remove a nest it is imperative to do it on very cool nights when the wasps become rather inactive. Caution is the watchword, though, because these are mothers defending their young and the stings can be painful and numerous, the sting not being barbed as with honey bees. Most nests are horizontal, and up to four inches or so in diameter. One, the nest of *Polistes exclamans,* starts out horizontal and droops to vertical. These nests become oval and sometimes more than six inches long.

Because the wasps are very active predators, they are certainly beneficial in the garden if they have nests in places where they can be tolerated.

POTTER WASPS

(Eumenes spp.)

Potter Wasp
Two times actual size.

It is the nest of potter wasps that make them conspicuous. Few would recognize the wasps, which look like smaller and slightly more knobby dark versions of paper wasps. These are solitary wasps, each female constructing nests and provisioning them for her own offspring. Each nest looks like a small jug, about half an inch in diameter, with a short sealed neck. When the female decides to make a cell, she selects a sheltered place, then carries dollops of mud there for the construction. This is a precision process, with a thin-walled pot resulting. When the pot is almost completed, with just room for her to get her head in, she starts to provision the cell with hairless caterpillars, which she has paralyzed by stinging them in the central nervous system.

Once the cell is full she lays an egg on the prey and restarts the cell-making process. She adds mud to the edges of the nearly spherical pot. Closing the sphere presents problems that are solved by simply adding extra mud and leaving a small neck. Try making a hollow sphere out of mud or clay and see whether you can seal it without making a neck, using just fingers.

The larva that hatches from the egg eats the prey, spins a cocoon inside the pot and pupates. When the new adult is ready to leave the pot, it simply makes a hole in the side and leaves. Using the neck would seem logical, but this is where the pot is thickest.

The females have stingers, but are very docile. About the only annoying thing about them is that they sometimes select unhandy places for their pots, such as inside thermostats.

Insects in Garden and Landscape Plants

While most of us do not attempt to grow valuable vegetables in our gardens, we do take them very seriously. The insects that feed on them are therefore of particular concern. Fortunately, there are not many species that cause problems, and most of the problems are easily taken care of. The spores of *Bacillus thuringiensis* are the first thing I try against any kind of caterpillar. The disease that results from the infection makes the caterpillars lose their appetites, and the treatment may provide protection for several generations. If conventional pesticides are resorted to, it is very important that all directions be followed closely to avoid leaving a residue on the edible portion of the plant.

AGAVE WEEVIL

(Scyphophorus acupunctatus)

Agave Weevil
Two times actual size.

The agave weevil contributes to destruction of more agave plants than any other single cause. The adults are dull-black beetles slightly more than an inch long, with the long, curved snout that is characteristic of the weevil group. They do not have functional wings, so have to disperse by walking.

When the adults feed, they do so with ordinary chewing mouthparts at the end of the beak. Their food is always agaves, although a closely related species attacks yuccas in southern California. The adult feeding is inconsequential, but female beetles deposit their eggs into feeding holes and the larvae that result eat into the plant, seeking the tender leaves and growing point at the center of the plant. They look like white grubs but lack legs.

Infested plants are very likely to be attacked by bacteria and break down into a slippery, smelly mass of tissue and woody fibers. The leaves appear to wilt and get wrinkled. If these symptoms appear it is too late for this plant to survive. If this mass is poked into, there will probably be larvae of other insects there also, eating the decaying tissue or the other larvae. When ready to pupate, the larva makes a cocoon of sorts out of woody fibers. One generation per year seems normal.

Eliminating the weevils from a neighborhood takes diligence. Infested plants should be completely removed, including the underground parts of the crown, because they probably hold pupae for the next generation. The largest species are all affected, but the common century plant, *Agave americana,* is the one most often seen in trouble. Wild agaves become susceptible when they start producing a flowering stalk, the time when the plant will soon die anyway.

This insect is a major pest of the maguey, used in Mexico as the source for sisal, pulque, mescal and tequila. The maguey plant has been introduced into other parts of the world as a crop, unfortunately with the weevil as a hitchhiker. The worm in the mescal bottle represents a caterpillar that bores into the base of the leaves, not this weevil larva.

ANTS

(Order Hymenoptera)

One could call these *wingless wasps* and be quite correct, for ants are close relatives of wasps and bees. Ants are studied by many because of their social behaviors. Scientists are fascinated with their organizations and how such large numbers of individuals get along so well. It is also intriguing to note that ant nests, except for one or two special times a year, are totally female.

There are many types of ants, from predators to farmers to dairywomen. Depending on their behaviors, they may need a stinger to capture prey, big mandibles to crack open seeds, or chemical defenses other than stings to drive away enemies. Thus one may be stung or bitten, depending on the type of ant one infuriates. Be

careful when standing or sitting because ants protect their nest sites quite aggressively.

Ants are the primary soil workers in the Southwest. Many people, when asked what is the most important organism for the soil, usually respond with earthworms. In desert caliche, I defy one to find earthworms, but ants abound by the millions. Their nest-making thus aerates the soil, helps retain soil moisture or drainage, and fertilizes and distributes many native plants. Next time you are cursing ants for being at your picnic, remember those trees providing shade needed their help first.

HARVESTER ANTS

(Pogonomyrmex spp.)

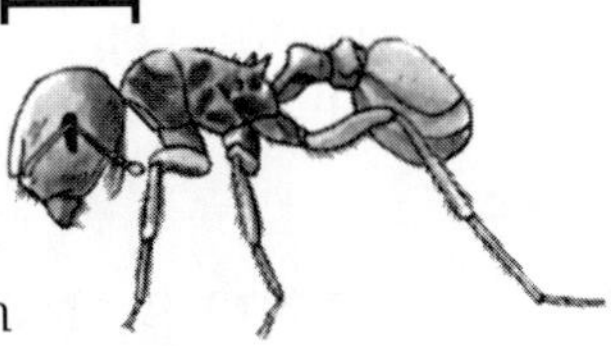

Harvester Ant
Four times actual size.

The stinging capabilities of the harvester ants, pogos to the entomologist (from their scientific name *Pogonomyrmex*), exceed even the southern fire ants in causing extreme pain and discomfort. The most obvious harvester ants are the black harvesters, ants about three eighths inch long. They make themselves conspicuous by clearing all plants from a circular area around the nest entrance. This is a common ant in the desert areas; higher up a close relative that is bright reddish brown does the same thing. Higher still another makes a large mound instead of a clearing. Another species in the desert always makes a small crescent-shaped mound just in one direction from the nest entrance. All of these ants are closely related. A feature they all share is a basket of long setae on the underside of the head that is useful in carrying soil as they dig the nest.

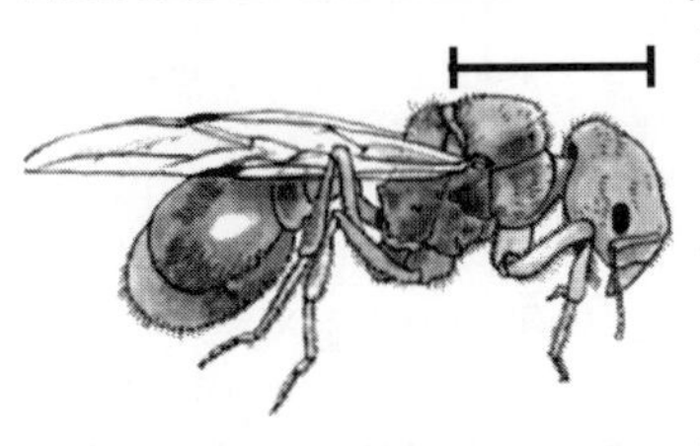

Harvester Ant Alate (winged)
Four times actual size.

The harvester ants that live in a clearing are aggressive in removing any plants that grow there. A shrub planted too closely might get cut up badly. Colonies can be very large. One dug up by some graduate students at the University of Arizona had 20,000 workers in it, and extended at least ten feet down. All of the passages and chambers were below the clearing on the surface, apparently maintained as a way of keeping plants from removing the moisture in the soil there. These ants, and most desert ants, show no interest at all in drinking water. They depend on being underground most of the time, where the humidity is high. The queen, larvae and pupae never come to the surface. Colonies can become many years old, as long as the queen still lives.

Harvester ants eat insects, but they also eat seeds, and actively gather seeds for food storage underground. Large numbers of their colonies usually are an indication of high seed production on range land. They can be a nuisance when reseeding efforts are undertaken. In the yard their main damage is in the presence of the nests and in the aggression of workers at the nest entrance. Even a small colony under a clothes line is not welcome. The sting is quite painful. It can induce shock in sensitive people or leave a large, red welt.

LEAFCUTTER ANTS

(Acromyrmex versicolor)

Leafcutter Ant
Four times actual size.

This ant is much more conspicuous for its activity than its appearance. The most obvious thing about them is a group of very symmetrical mounds of particles of fresh soil, more or less like volcanic cinder cones that may spring up in a short period following summer rains. Some of these may be as much as a foot high and be visible for some distance. The worker ants that make the cinder cones are reddish brown, from three sixteenths to three eighths of an inch long, and not at all shiny. If one observes them closely, they have many spines on the body and a triangular head.

Somewhere in the vicinity of the cinder cones there will be one hole that has some pieces of fresh leaves lying around it. This is the main entrance to the colony. All of the other cones are the result of workers carrying soil straight up from their excavations and depositing it at the lip of the cone. It is a very efficient system for an ant that makes a nest that has a broad lateral dimension. Workers forage out from the main entrance in search of leaf material, usually at night during hot weather, but during the day when the days are cooler. They pick up freshly fallen leaves and leaflets, or cut fresh leaves from many kinds of plants. Many a homeowner has come out in the morning to find their bougainvillea or other favorite plant naked after a night's work by these ants. The leaf material is carried back to the nest and taken inside. If the leaves are dry enough they stay down but, if they are still damp, they tend to get brought back up for a period of drying at the top. The search for leaves seems to depend on the demand for food downstairs, because nests may remain closed up and appear abandoned for weeks at a time. The underground portion of the nest is very large, and usually extends below the caliche layer in the desert.

What happens to the leaves is most remarkable. They are chewed up by the workers and used as a mulch in the cultivation of a special fungus, which was brought to the colony by the founding queen. The ants are able to suppress the growth of other fungi, and depend entirely on the one they grow. Several related species, in the American Tropics, are larger, have larger colonies, and may make agriculture almost impossible where they are common. These are known as *parasol ants* from their appearance when carrying large pieces of leaves.

SOUTHERN FIRE ANT

(Solenopsis xyloni)

New residents to the Southwest from the Midwest or Northeast are likely to have a firmly held conviction that ants bite and don't sting. Their first encounter with the southern fire ant should dispel this idea. The sting of these ants is impressive.

The worker ants themselves are distinctive, varying from about an eighth to a quarter inch in length, shiny, and two-toned bright reddish tan and black. Colonies are likely to be in the dampest part of a yard. A well-kept lawn is ideal. Gardens also serve. Individual ants are not very aggressive, but any disturbance of the entrance to the colony brings immediate reprisal.

For most of the year the colony is actually quite scattered under numerous small mounds over many square yards. A person sitting on one of the entrances is very likely to get stung. When the colony

is ready to release flying ants in the summer, it comes together and constructs a formless mound as much as a foot across and six inches high. Then the nest guards really become aggressive.

Small children playing on the grass are the most frequent victims. Parents should be alert to the presence of the ants. Poison bait seems to be the most effective control at present. If it weren't for their nasty disposition, fire ants would be regarded as beneficial predators. They are extremely diligent in killing insects, putting their stingers to good purpose.

APHIDS

(Family Aphididae)

Aphids or plant lice are known to all gardeners. They come in a variety of sizes, from one to three sixteenths of an inch as adults. Their color ranges from green to bright yellow and black, or brown. They form small colonies on the plants they infest, colonies that generate sticky honeydew and may extract enough from the plant to cause curling of leaves and even death.

At lower elevations aphids are almost entirely a problem of the cool spring season. Later their numbers are cut down drastically by lady beetles and other predators and parasites. Each colony on a plant is started by a single female, which produces living young without benefit of sex in a process called *parthenogenetic* ovoviparity. The original aphid is winged but the first offspring become wingless adults that soon start producing young of their own. When the colony gets crowded, the young develop into winged adults, which go through a flying period before settling down on a new plant.

The almost universal presence of aphids on plants conceals the fact that most of them are very plant-specific. The yellow-and-black aphid on oleander can live on this plant and milkweeds, nothing else. The brown aphids that make a shiny mess under arbor vitae trees can't even live on the related junipers. A few species have a wide host range. Some species are so prolific that aphid problems are assured if their food plants are present. The common aphids on cole crops and cucurbits such as cucumbers fit this category.

Dealing with aphids in the spring almost inevitably involves the application of a pesticide with a short residual period if an edible

plant is involved. For roses and other ornamental plants it is possible to use one of the systemic insecticides that is moved around in the plant and very deleterious to populations of sucking insects. If the infestation is not causing serious damage to the plant, it is worthwhile waiting a while. Many years parasitic wasps reduce the populations by themselves, and they are helped immensely by lady beetles as soon as the weather warms.

BAGWORMS

(Oiketicus toumeyi)

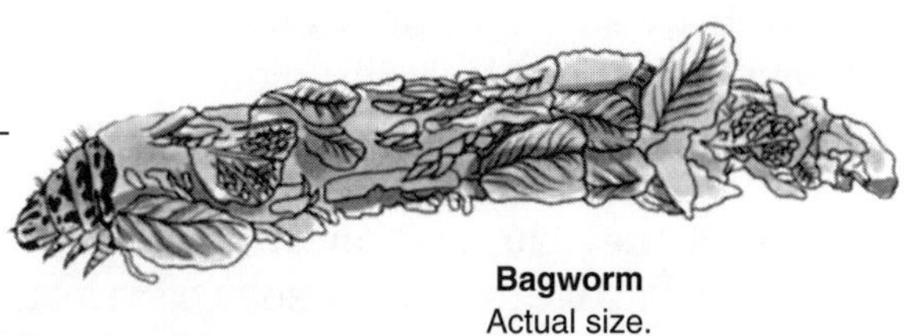

Bagworm
Actual size.

Slender, tapering pale-tan cases, as much as three inches long, and containing a naked brown caterpillar, are bagworms. The cases are made of leaves held together with very tough silk. The leaves soon die and turn brown. The cases can be found on a very wide variety of trees and shrubs, ranging from conifers to eucalyptus, which otherwise is almost insect-free in North America.

The moth that goes with the caterpillar is rarely seen. It is about an inch long, brown and slender with almost transparent wings. The flying moths are all males. The adult female is a legless, caterpillar-like individual. It stays inside the case after emerging from the pupa until a male finds and mates with her. After mating, the female almost fills the case with eggs and dies.

The hatchlings spread out from the case. Dispersal of bagworms depends on the hatchlings getting out on a twig and letting out a long strand of silk. Even a gentle breeze will carry them away, and to a possible new source of food. Once the young caterpillar has constructed a case, the chances of dispersing from the home plant are decreased.

When not feeding, and particularly when they are ready to molt or pupate, the caterpillars attach the cases to a twig with silk. It is very durable silk, and cases may be on the tree for a decade before weathering takes them away. Heavy infestations are a rarity, probably because there is a common parasitic wasp that attacks the caterpillars. Our common bagworm species is *Oiketicus toumeyi,* originally sent to a taxonomist by Professor Toumey, the first botanist at the University of Arizona. Another species, *Thyridopteryx meadi,* specializes on creosote bush.

CACTUS LONGHORN BEETLE

(Moneilema gigas)

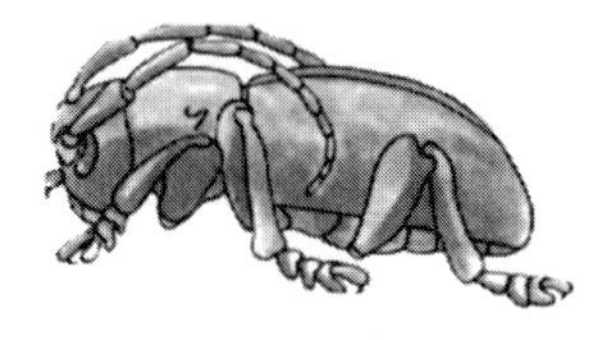

Cactus Longhorn Beetle
Actual size.

The cactus longhorn beetle is the primary pruner and reshaper of cacti of all kinds in the Southwest. The beetles are quite attractive, shiny black with some white markings on the long antennae, and very hard-bodied. They can be found at any season of the year, and always show an interest in cacti. Growers of exotic cacti loathe this beetle.

The feeding of the adults is confined to succulent tissue. Certain species of cacti are preferred above all others, and serve as a reservoir of the beetles. Really spiny cacti are usually safe from adult feeding, but many chollas and prickly pears aren't spiny enough. Feeding on the terminal portion of barrel cacti can result in the cactus stopping growth, eventually putting out side heads, but perhaps five years after the fact.

Females lay single eggs in many different cacti, even ones the adults don't use as food. The larvae are pale grubs with brown heads that burrow into the cactus and feed, usually pushing semi-liquid frass from the burrow entrance, green at first but turning black. Some may go down into major roots to feed, showing no frass on the surface. The large Burbank prickly pears seem capable of supporting a large number of larvae without showing damage, but some other cacti die and melt down rapidly. Young saguaros up to six inches high are susceptible.

Control consists of a search-and-destroy mission in early summer, or at any time that adults are found. Early in the morning and late in the afternoon seem the best times. A warm rain really gets them moving. Because the adults have no flying wings, they are easy to catch unless they get too deep in the cactus tangle. Once a population has been reduced in a small area, it takes quite a while for new beetles to walk in and get things going again.

CATERPILLARS
(Order Lepidoptera)

When one refers to worms, this is probably the stage of insect people think about, caterpillars. This is the eating machine which sometimes defoliates a plant or leaves a plant looking unsightly. This is the insect that drops all those black pellets around the base of a plant, fertilizing the soil for that plant. This is the stage which causes the most consternation on the part of the gardener or landscaper, because they don't cooperate with what people deem aesthetically pleasing. The caterpillars do not always cooperate with our gardening practices either.

What they eventually change into, though, are many of the night-flying moths which supply the nocturnal creatures with food or delight people with their appearance at the porch light.

Caterpillars don't set out to kill your plants, for then what would the next generation have to feed upon? Death may occur to a plant, but it probably was doomed before the caterpillars appeared. Normally a healthy plant and insects get along just fine in their give-and-take world.

CUTWORMS AND LOOPERS
(Family Noctuidae)

Probably because some of the parasitic insects that keep them in check during the summer do not thrive in cool weather, several kinds of broad-spectrum caterpillars can become bad pests of gardens grown during the winter and spring at lower elevations. Relatively few species of moths and butterflies have caterpillars that are able to feed on a variety of plants. Two moth caterpillars are the ones most often involved in damage to plants during this period.

One of these is the granulate cutworm, *Feltia subterranea,* a mottled-brown caterpillar that feeds well up on the plant despite the cutworm part of its name (which implies that it cuts plants off at ground level). The other is the cabbage looper, *Trichoplusia ni,* a green caterpillar that lacks abdominal prolegs except at the tip of the abdomen, so moves by looping along rather than crawling. The adult of the cabbage looper is easily recognized at lights by the

presence of a silver "Y" on its front wings. The adult granulate cutworm is one of numerous mottled-brown moths at the light.

Caterpillars are particularly susceptible to infection by *Bacillus thuringiensis,* which can be introduced as spores. The standard commercial source contains chemicals produced by the bacillus that reduce the appetite of the caterpillars even before the bacteria have reproduced.

SOUTHWESTERN SQUASH VINE BORER

(Melittia calabaza)

Squash Vine Borer
Actual size.

One of the most frustrating experiences for a summer gardener is to have a promising stand of summer squash suddenly wilt and die, just before the first crop is ready. This is the work of the squash vine borer in the Southwest.

The culprits are brown-headed white caterpillars that bore into the main stems of the squash vines. They are over an inch long when fully grown. The eggs from which they hatched can usually be found on the infested stems. These are of pinhead size, tan, and very shiny. When the caterpillar has completed its feeding, it leaves the stem to spin a loose cocoon and pupate. The emerging adult looks more like a flashy wasp than a moth. It is an inch long, with transparent hind wings and hind legs strikingly marked with a thick covering of orange-and-black hairs. The moths are diurnal, and can be seen loitering suspiciously around any uninfested plants.

What we have is the Southwestern squash vine borer, which ranges into southeastern Arizona from Mexico. Its existence was discovered only a few years ago. Until then there was an unexplained discrepancy between the squash varieties infested in the Midwest and the Southwest.

Some have suggested slitting the squash stems and killing the caterpillars, then covering the stem with soil to encourage growth of lateral roots. I have never been very successful at this, and have resorted by buying squash from areas where they are easy to grow.

Wild gourds are the native host of our borer, and the adults seem to be able to find every squash vine.

TOMATO HORNWORM

(Manduca quinquemaculata)

The official common name of the big caterpillars we find eating our tomato plants, as recognized by the Entomological Society of America, is *tobacco hornworm, Manduca sexta*. The tomato hornworm is a different species, not as common to the Southwest. Both species eat tomatoes, peppers, egg plant, tobacco, Jimson weeds and other succulents in the plant family Solanaceae.

The caterpillars grow to more than three inches long and half an inch in diameter. They are green with oblique pale bands across the sides of the body. The body ends with a stiff pointed horn a half inch long, decoration only, not a stinger as many presume. The camouflage of these caterpillars on a tomato plant is so perfect that the best way to find them is to look for the large, black fecal pellets under the plant. Even when their current place of activity is discovered, finding them takes sharp eyes.

After the caterpillar stops feeding, it leaves the host plant and finds a soft place to dig deep into the soil. After burrowing several inches below the surface, the caterpillar forms an earthen cell with saliva and soil, a pottery cocoon if you will, and changes exoskeletons. This new look is a dark-brown, elongated form with a jug-handle near the head region. If examined closely, imprints of the wings, legs, eyes and antennae can be seen, and the jug-handle is the proboscis or tongue of the new adult moth. This stage is immobile, but it may wiggle its rear end if disturbed. This bizzare form is the pupal stage and inside the remarkable transformation from worm-shape to adult moth is occurring.

The moth that goes with the caterpillar can be characterized as a rapidly flying gray cigar, with some lateral yellow marks on the abdomen, and about two inches long. These are sometimes seen visiting blossoms at dusk, hovering in front of the flower and extending a long proboscis or tongue into the nectary. They may not be particularly abundant, but the females spend many night hours searching out food plants on which to deposit a single egg.

Parasitic flies and wasps have a great deal to do with the fact that these caterpillars are not more abundant. One small wasp, a species of *Apanteles,* sometimes puts ornaments on the caterpillar in the form of numerous white cocoons the size of a grain of rice. The female wasp inserts a single egg into the caterpillar. This egg undergoes repeated cell divisions before hundreds of embryos are produced. The resulting larvae eat the internal contents of the caterpillar, avoiding the vitals, and finally come to the surface to spin cocoons and pupate. A heavily parasitized caterpillar may continue to feed normally, but the overall effect on the population of hornworms is a reduction. Hand-picking is the usual control in a small home garden.

WESTERN GRAPE LEAF SKELETONIZER

(Harrisinia brillians)

The grape leaf skeletonizer is a good example of a man-made pest. It is native throughout the Southwest and into Mexico, feeding on wild grape in numerous canyons. It is a very scarce insect there because several parasitic insects keep its populations at a low level. It can become a very serious pest of grapes in urban plantings. The individual gardener may not have been responsible, but the parasites have a hard time competing when pesticides are used regularly.

Grape Leaf Skeletonizer
Actual size.

The gardener becomes aware of the skeletonizer by discovering that big patches turn brown on the leaves, only veins remaining. They are the result of feeding by rows of small caterpillars, perhaps a quarter inch long, black and yellow-banded. Later stages of the caterpillar go off on their own, consuming whole leaves, not just one epidermis and the parenchyma as the young ones do. They grow rapidly, ending up leaving the plant to spin loose cocoons and pupating. Adults emerge from the cocoons in a week or so and the cycle is complete. Unchecked, the caterpillars can defoliate the vines. The moth looks much more like a blue-black wasp, slender and about an inch long. The moths are day-flyers and obviously interested in the grapes. Females lay masses of eggs on the backside of grape leaves, which hatch into the little wrecking crews first seen.

Despite their small size and delicate appearance, the caterpillars have erect venom-bearing hairs all over the body. Careless handling can cause a mild burning sensation to tender skin areas touched by the hairs. Because the hairs are numerous and break off easily, it is a good idea to avoid really shaking things up in infested vines, lest spines end up floating in the air and getting into your eyes and lungs.

The caterpillars are vulnerable to the standard pesticides, but are most effectively controlled by the application of *Bacillus thuringiensis* spores, which seem to be more effective on this species than on some others. One application often gives protection for a whole season.

WESTERN TENT CATERPILLARS

(Malacosoma californicum)

Throughout the Southwest, wherever there is enough water for willows and cottonwoods, the activities of the tent caterpillar may be in evidence. The caterpillars appear soon after the trees have leafed out in the spring, making large gray webs as much as a foot and a half long and almost as wide. The webs are a group effort of sibling caterpillars, all of which hatched from a single egg mass. They stay together in the web during the day and forage out to feed on the young leaves at night. The caterpillars are almost black, up to an inch and a half long, and sparsely covered with long hairs in a variegated color pattern.

When they have completed their development they leave the communal web and wander around until they are ready to pupate. They spin a loose cocoon with white chalky material on its surface. Then they pupate and emerge in a couple of weeks as weak-flying rich tan moths three quarters of an inch long. The moths soon mate, and the females lay masses of eggs on the host trees. These eggs cannot hatch until the following spring.

Tent-caterpillar populations have wild fluctuations from year to year. There are parasitic insects that affect all stages. There also is a very contagious disease that is capable of taking the numbers close to zero. One of the most effective ways of handling the problem in a region is to import sick caterpillars from another location. The contagion seems to be wind-borne. The trees readily recover from the loss of that first set of leaves, producing a second set of probably less-edible leaves soon after the caterpillars have finished feeding. The tent caterpillar isn't very welcome if the tree is in a patio or over a pool. Some of the droppings stay in the webs, but many make it to the ground.

Tents seen on wild cherry during midsummer are those of another caterpillar, the fall webworm. The life cycle is very much like that of the tent caterpillar, but on a different schedule.

COCHINEAL
(Dactylopius confusus)

Cochineal belong to a group of insects called *scales.* This group normally appears as flat, cryptically colored insects that do not move about on the host plant. They have a beak that penetrates the plant and allows feeding on the plant juices. Many scales exude a crusty material over the soft body that protects them. The cochineal secretes white waxy materials which protect them from most enemies and the environment. This white wax makes them resemble the mealybugs, a close relative of scales.

Cochineal is readily seen on prickly pear pads and cholla canes in urban scenes resembling a spitwad war zone. The actual insect is seldom seen though, unless one teases away the wax to expose a quarter-inch-long red bag, a female cochineal. This animal may remind one of the engorged dog tick. This red pigment, an anthraquinone, seems to protect the cochineal from most predators. This may also be the key to stopping parasites from attacking, as this insect is one of few with no known parasites affecting it. One predator that is not deterred is the caterpillar stage of a small moth, *Laetilia coccidivora.* This is a strange occurrence because few moth larvae are carnivorous. A small ladybird beetle is also a common predator found in the waxy home of the cochineal.

The cochineal disperses in a stage called the *crawler.* These juveniles hatched from eggs laid underneath the female, moved to a feeding spot and produced long wax filaments, then moved to the edge of the prickly pear pad where the wind catches the wax filiments and carries the cochineal to a new host. These individuals establish feeding sites on the new host and produce a dense wax covering surrounding the individual and others that have settled nearby. There are male and female cochineal insects, but the males

are very tiny and have wings that allow them to fly about in search of mates on different cacti. Males are seldom observed.

Cochineal, which means *scarlet-colored,* is famous as a dye in both the textile and food industries. When Cortéz landed in the New World in 1518 and proceeded to conquer the Aztecs led by King Montezuma, he discovered a highly developed textile industry with brilliant red garments. The Aztecs produced the red dye from an insect they called *nochezli* that fed on a cactus called *nopal.* The Spaniards eventually sent bags of dried cochineal back to Spain and the red dye use spread into many countries.

Michelangelo bought it to use in paintings, the British 'redcoats' and the Canadian Mounted Police coats were dyed with cochineal red. It is thought that the first U.S. flag made by Betsy Ross had cochineal red stripes. Other famous fabrics reported to have been dyed with cochineal include the breeches of the Hungarian Hussars, the Turks' Fez and the skull caps of the Greeks. Cochineal dye replaced another scarlet dye obtained from the insect called *kermes,* a scale insect that feeds on oak trees.

Modern use of the cochineal dye is limited, but has some followers in the craft world who weave and dye their own fabrics. There is also some use of cochineal as a food coloring. Because it is of insect origin some people might be squeamish about such use. Several other less-expensive aniline dyes have turned out to have deleterious effects in food so cochineal red dye may make a comeback. Before aniline dyes, cochineal was in such demand that the insect and the prickly pears that supported them were introduced into suitable climates throughout the world. Only a few of these areas, Oaxaca, Mexico, Algeria and the Canary Islands, were successful and now produce most of the dye used in commerce.

Extremely heavily infested prickly-pear pads may die, but this plant is so prolific that removing damaged pads should take care of the problem. A strong stream of water aimed at the waxy cochineal should clean them up aesthetically.

CYPRESS BARK BEETLE

(Phloeosinus cristatus)

One of the most interesting conifers cultivated in Southwestern cities is the Arizona cypress. It is a native tree of damp mountain

canyons in the southeastern fourth of Arizona. Unfortunately, almost every tree suffers water stress in cultivation, with the result that very few mature trees are to be seen. The extravagantly watered campus of the University of Arizona has some beauties. Few other places do. The tight red bark is spectacular.

The trees are done in by the activities of the cypress bark beetle, a native insect to go with the native tree. The first sign of an infestation is the tips of branches turn almost white. If these are examined, one finds a hole at the base of the dead portion, and in the hole a pair of small, reddish-brown oval beetles less than a quarter inch long. You have disturbed them in their nuptial chamber, which they have excavated cooperatively. Their housing also serves as their food. The next step is for them to leave this chamber and progress to a main branch or the trunk, where they excavate a tunnel several inches long in the cambium, parallel to the trunk. The female then goes along the tunnel and inserts eggs at close intervals. The larvae that hatch from the eggs make tunnels perpendicular to their parents' tunnels. These also are several inches long. An amazingly few pairs of beetles can produce larvae that kill the bark all the way around the tree, effectively girdling it.

FIG BEETLE

(Cotinus mutabilis)

Large green beetles with yellow margins are fig beetles, a good inch long and chunky. They make an impressive addition to the landscape at lower elevations in July and August. When they fly they show off iridescent purple-black wings. At close range one can see that the underside is a metallic green and the upperside a delicate matte.

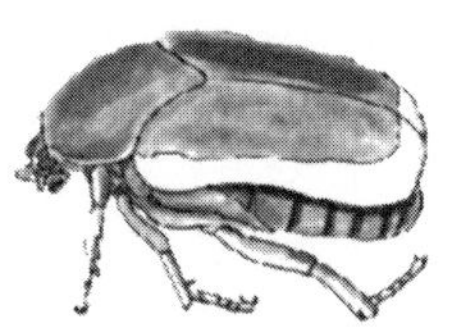

Fig Beetle
Actual size.

The head has a short square horn. Antennae are present but they are the specialized short, leaflike antennae of scarab beetles and kept out of sight under the head much of the time. Fig beetles are scarabs but are an exception to the normal nocturnal activity of this family.

If it weren't for the fact that these beetles have a sweet tooth to indulge, they would be welcome in the garden. Children can tie

threads to them and make tethered aircraft. They are very durable. They love ripe fruit, and may descend on a ripe fig, peach or apricot in such numbers as to hide the fruit entirely. They leave nothing to pick. The gardener who is growing the fruit in order to have it tree-ripened does not appreciate fig beetles. They are even more diligent than the birds. Bagging the fruit before it is ripe is about the only feasible protection. After feasting on the many cultivated desserts in the area, the fig beetle may go to its natural food, prickly pear fruit, for dinner.

Fig Beetle Larva
Actual size.

The larva of a fig beetle is a large white grub, at least two inches long. It differs from most white grubs in not taking a C-shaped appearance, plus it flips over on its back and walks away upside down when it ends up on the surface. It also is hairier than the standard scarab-beetle grubs. The larvae are usually found in gardens because they depend on a very highly organic environment. Heavy mulching may bring them in. Their favorite home is compost of all kinds. They are beneficial there, eating the material and passing numerous fecal pellets. When they are ready to pupate, they form a hardened shell of soil and pupate inside it. The normal life cycle takes a year, so the pupa is the usual overwintering stage.

The fig beetle and its eastern cousin, the green June Beetle, *Cotinis nitida,* have similar larval behaviors. In a Cherokee myth its behavior has been explained as follows: Acting as Chief of a council meeting, "Grubworm" became so amused at a suggestion that he shouted with glee and fell over backwards in his excitement. When Grubworm tried to get to his feet he could not and had to wriggle off on his head. Thus today the larvae of the genus Cotinus are known for moving on their backs unlike most other white grubs.

LEAF-FOOTED PLANT BUGS

(Leptoglossus zonatus)

Leaf-footed plant bugs have incomplete metamorphosis and long sucking mouthparts that they use to reach and feed on developing

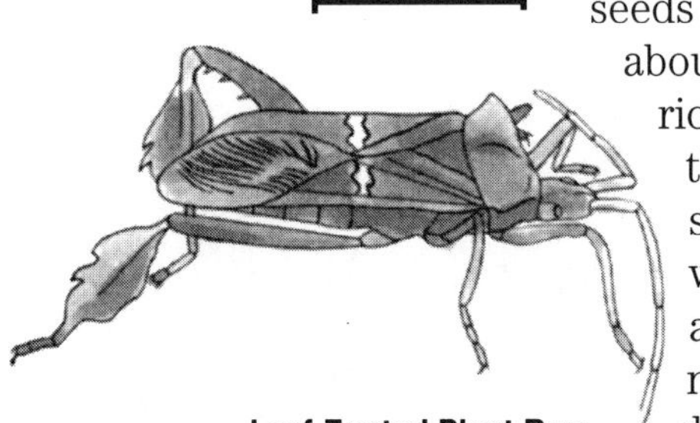
Leaf-Footed Plant Bug
Twice actual size.

seeds in a variety of plants. Adults are about three quarters of an inch long, rich brown with a pattern of yellow that varies slightly between species. The immatures are somewhat redder, have black markings and lack the pale markings. Both nymphs and adults have well-developed scent glands, from which they can eject a strong-smelling liquid.

Because they have to have seeds that have not yet developed a hard coat, these insects have a rather restricted food plant range. Plants with very small seeds are immune. The main garden fruits damaged by their feeding are pomegranates, citrus and pecans. They can also make a living on yucca and agave pods, but these aren't something we harvest. In a neighborhood the infected plant is almost always the pomegranate. It produces fruit over a long period and has numerous seeds that remain soft for a long time before they mature. Adult bugs fly from there to the other hosts. Pecans aren't susceptible for very long. Citrus is. Abortion of the seed seriously damages pecans. In citrus the big damage is through the introduction of bacteria, often resulting in a hole and decay, for which not entirely blameless woodpeckers are blamed.

Management of the pomegranates will alleviate the problems with other fruits. Most pomegranate hedges are left untended. Removal of fruit for a period should reduce bug populations; the nymphs won't be able to walk very far.

MITES
(Family Eriophyidae)

These relatives of ticks and spiders are the no-see-ums of the plant world. They have the ability to suck plant juices and in so doing create a new look for the plant. One will not see these mites with the naked eye, but the results of their feeding are unmistakable. These are Nature's artists at work.

ALOE GALL MITE

(Eriophyes aloinis)

All gall mites are tiny enough so they are barely visible on a contrastingly colored surface. They are recognized by the things they do to plants. All of them have close associations with a small group of plants, sometimes a single species.

The aloe gall mite lives on several species of the genus *Aloe* from South Africa. It forms small colonies on the plants, and everywhere it feeds the plant reacts by producing tissue that is not typical for the plant. On the leaves this is in the form of yellow canker-like growths. On the flower stalks the usual manifestation is a flattened and twisted deformation. *Aloe* fanciers don't appreciate the deformation, but dried-flower arrangers may use them as centerpieces.

The *Aloe* species most often seen with heavy infestation of the aloe gall mite is the common soap aloe, *A. saponaria.* This is widely planted at lower elevations. This is the species with bands of pale spots on the leaves and an open spike of orange blossoms. It offsets freely, so most plantings end up as beds. The mite is a pest in South Africa, and seems to have been introduced into North America fairly early.

Control consists of removal of the affected plants. If this is done diligently, even a heavy infestation can be eliminated. It will probably have to be done repeatedly to be complete.

PALO VERDE GALL MITE

(Aculus cercidii)

The palo verde gall mite is as tiny as other gall mites and not conveniently visible. What it causes the palo verde tree to do is quite visible at the speed limit on the interstate. Clusters of small branches, usually darker green than normal, appear on the outer parts of the tree, where growth is most active. At a quick glance they look a lot like mistletoe, and are often called *witch's broom.* Over a period of several years these are likely to become larger and increase in number, with infested branches finally turning brown and dying.

The plants affected so far seem to be confined to the genus. Large blue palo verde trees were infested on the University of Arizona campus at least 20 years ago. The landscaping of I-10 near

Tucson resulted in a row of trees on the median exhibiting strong symptoms. More recently palo brea trees, an import from Sonora, have also become infested.

Even microscopic examination of the infestation may not reveal the mites. They seem to be active and feeding only on young, actively growing terminal buds. Their feeding is what produces the symptoms.

As is the case with other gall mites, control is best by elimination of the infested parts of the plant. Diligent pruning is in order. Infestations should be watched for particularly on trees being pushed to fill in a landscape function. Active growth is what favors the mite. Established trees that are not given supplemental watering may not show much effect.

PALO VERDE ROOT BORER

(Derobrachus geminatus)

Palo Verde Root Borer Adult
Half actual size.

The palo verde root borer in the larval stage is grublike and up to five inches long. The adult beetle, a cerambycid, is three to three and a half inches long with long antennae, short stout mandibles and a sharp spiny pronotum (the plate behind the head). The adult beetles are active in the summertime, flying mainly in the early evening hours, becoming a hazard to navigation for motorcyclists.

The larvae consume the roots of trees, boring into the starchy middle area although sometimes cutting across and removing the entire root. The larvae hatch from eggs deposited in the soil under the tree by female beetles. Infested trees seem to have three sizes of larvae present most of the time, so a three-year life cycle is assumed. The palo verde part of the common name is a little unfortunate, because the larvae affect a variety of non-native trees at lower elevations. The name came from the almost universal presence of larvae on the roots of dead Mexican palo verde trees in Tucson.

Root Borer Larva
Half actual size.

Emerging adults make a smooth-sided hole under the infested tree. This isn't an access hole for the female to deposit eggs, so pouring pesticide down the hole isn't particularly useful. Perhaps it would soak into the surrounding soil and kill larvae still there, perhaps not. As with other consumers of trees and shrubs, an investment in water and fertilizer would probably be more appropriate. Such care would help the tree outgrow the damage.

PALO VERDE WEBBERS

(Bryotropha inaequalis)

This is yet another insect that is quite inconspicuous alone, but whose activities make it conspicuous at the landscape level. The webber is a small caterpillar, no more than half an inch long and slender. What it does is to build itself silken tubes, in which it lives. The normal food plant is the foothills palo verde, *Cercidium microphyllum*. The webber appears soon after the palo verde trees leaf out in the spring, and may be present in such abundance that the aspect of the trees is that they are covered with a wispy layer of white something. Whitethorn, *Acacia constricta*, may also be involved in the webbing.

Because the palo verde tree is very resilient, and drops its leaves whenever there is moisture stress anyway, there is no damage done. Should a palo verde owner want something to worry about, the caterpillars may do a little feeding on the bark of twigs, leaving permanent marks.

The moth that goes with the caterpillar is a tiny tan one that is as inconspicuous as its larva. All stages must contribute to the nutrition of insectivorous birds and lizards.

STEM-BORING BEETLES

(Order Coleoptera)

There are many niches available to insects, the inside of plants being very attractive to those with the right adaptations. Some groups of beetles are drawn to an unhealthy or dying plant to lay eggs in this wonderful nursery. Here the immature beetles, soft larval forms with strong mandibles, can feed out of sight of predators.

They are safe from the harsh environment until they can emerge as an adult beetle. You can see the results of many of these feeding frenzies by cutting a piece of wood and seeing the characteristic oblong or round galleries inside the branch. Once again, it is normally unfit plants that bear the brunt of these diners. Pruning and proper care of plants will deter most of these insects.

GIANT PALM BORER

(Family Bostrichidae)

(Dinapate wrighti)

Giant Palm Borer
Actual size.

As I write this I feel that it is predictable that the Southwest will be having a major problem growing palm trees in the next few years. The reason: the giant palm borer beetle has spread dramatically from its native home in *Washingtonia* palms in southern California into planted palm trees of both the *Washingtonia* and the *Phoenix* types. It was found in Phoenix, Arizona palms in the late 1980s and recently found on the golf-course palms of Las Vegas.

Adult beetles are huge, as large as most people's thumbs, brown and nearly cylindrical. Grubs that go with them look like oversized white grubs, quite flaccid when extracted from its gallery, but endowed with impressive mandibles for chewing through the fibrous innards of a palm trunk.

The life cycle of the larval borer appears to extend for three to nine years. When adults emerge from infested palm trunks they almost immediately seek out the crown of palms, where they bore into the terminal, mate and lay eggs. The grubs that hatch from the eggs begin eating down inside the trunk, going almost all of the way to the ground, then back up the tree. Because of the diffuse nature of the growing part of a palm trunk, the larvae do not cause immediate damage to the tree or its conducting tissue. Many grubs feeding inside the palm can weaken the trunk to the point that it may snap off in a high wind.

When the grubs have completed their feeding, they bore outwards and pupate near the edge of the trunk, facilitating easy emergence as adults. The emergence holes are about the size of a quarter, easily visible and a good indication of how infested the palm may be.

OCOTILLO BORER

(Family Buprestidae)
(Chrysobothris edwardsi)

The larvae that eat the cambium of ocotillo and outright kill this tough plant are closely related to the ones that bore into the cambium of roses and apple trees. However, they belong to a species of Buprestidae that has adapted to ocotillo and apparently can develop in no other plant. The boojum tree and several other related species have never been examined, but the species probably invades those plants as well.

The larvae get into the ocotillo through the adult female laying eggs into the bark. As seems to be the case with all wood-boring insects, the emission of certain compounds from stressed plants is the cue to the attraction. Insects generally have an acute sense of smell, but not on a broad-spectrum basis. Pheromones emitted by female moths are detected by males so far downwind that they must be reacting to single molecules and flying upwind in response to a very weak mixture. The parsley butterfly finds single plants of parsley and related plants among the many smells of the urban environment.

Rapid decline of ocotillo usually comes within the first year after they have been transplanted. The usual way of moving them is to chop off most of the roots and just move the cleaned and dry plant to the new location. The root chopping has to be stressful. If this is done during the period that the adult beetles are flying during the summer, the attractive plant will have picked up some larvae before it reaches the place where it is to be sold. An infestation is inevitable. Avoiding this susceptible period seems to be all it takes to avoid the problem.

Once the beetle is well established in a neighborhood, even slightly stressed plants may get some larvae. The easy way to find this out is to water the ocotillo during the warm part of the year. It should leaf out. Any branches that do not probably have larvae in them. Cut them off below where the leafless part starts and get them out of there. Piling them up after cutting them will not hamper the larvae at all. If the plant does not leaf out, dig it up and dispose of it.

ROSE CANE BORERS

(Family Buprestidae)

There is no insect with this official common name, but insects boring into the canes are a common concern of rose growers. These are almost always flat-headed wood borers, larvae of beetles of the family Buprestidae. The larvae are yellowish-white, legless, slender in back but with the front part of the body expanded laterally and flattened. Big ones are an inch long.

Several species are involved, and they also infest some other kinds of plants in the family Rosaceae, including apples. The way in which an infestation gets started is that the plant gets under stress in some way. The plant then produces a characteristic odor, which the female beetle homes in on. She then deposits her eggs, one at a time, into the bark of the plant. The larvae that hatch bore down into the cambium and feed on this nutritious layer. Because the plant depends on the cambium to move nutrients and make wood and bark, the results can be devastating. Girdling may result if the larva eats all of the way around the stem. Avoiding stress on the plant is important, particularly in the summer, when the beetles are active. Sunburned canes are particularly susceptible.

Several different kinds of small wasps and bees use the cut ends to gain access to the pith, which they dig out to make holes in which they can construct cells to provision with insect prey or pollen paste for their larvae. Loss of the pith may cause the tip of a cane to die. It is probably this fact that has led gardeners to believe that the borers result from not sealing up the ends when roses are pruned.

Once established, borers are very difficult to control. Prevention of stress and removal of infested canes should keep the population in check. The brown bark over the burrows of the larvae give away their location. They can be cut out with a sharp knife.

WHITE GRUBS

(Family Scarabaeidae)

White grubs fit their description, being C-shaped white insects with tan heads. They live in the soil and consume soil for its organic content and the roots of many plants. They are the larvae of scarab

beetles of many kinds, including the famous Japanese beetle, *Popillia japonica,* of the eastern U.S. Different species have different preferences, some confining their feeding to organic matter in high humus soils (as the case with fig beetle larvae), others really concentrating on roots. Plants weaken and pull out easily because they have lost most of their roots.

The main mystery is how the grubs get into the soil in the first place. They may appear in large numbers within a few weeks during the summer. The explanation is simple. Adult scarab beetles are almost all nocturnal, and spend the day in the soil. When the female is ready to lay eggs she does, wherever she happens to be. If she has been feeding on the leaves of a particular tree, the eggs and larvae will be under that tree. If she happens to belong to one of the species that is strongly attracted to light, she may end up in the soil under the light.

The highest populations of white grubs are found where the digging is easy. The combination of bright outside lights and friable soil under them is almost certain to generate a white-grub problem.

Poisonous and Venomous Creatures

We need not concern ourselves too much about the poisonous insects and other arthropods around us because we rarely have them on our diet. But there are some. Avoid anything that is brightly colored if you are reduced to entomophagy (eating insects). Lady beetles smell and taste bad. One or two blister beetles in the soup could do anybody in. They even kill horses. The monarch butterfly is famous for sequestering enough poison from the caterpillar's food to make the adult a real cup of hemlock to some birds.

Venomous is something else. Venom has to be delivered by some type of hypodermic device. This can be the fang of a spider or the stinger on the tip of the scorpion's abdomen. Beware the caterpillar with branched spines because those spines mean business. One though, the puss caterpillar, looks very soft and cuddly, a wolf in sheep's clothing. The soft fur conceals some erect poison spines that can deliver a painful dose of venom. Almost everybody instinctively avoids wasps and bees, but might not recognize the brilliantly colored velvet ants as wingless wasps.

Simply avoiding the venomous is the logical way to go. Don't put hands under things in picking them up, and watch the ground when you put your hand down. Few people are stung and many fewer have any violent reactions.

The reader is referred to *Venomous Animals of Arizona* (1982) by Robert L. Smith. It is a treatise dealing with venomous animals of Arizona with Southwestern distribution. This source also has information on many of the creatures called *fearsome but harmless* in this book.

CATERPILLARS WITH STINGING SPINES

(Order Lepidoptera)

Most of the time people tend to regard caterpillars as just some ugly, fleshy worm. The caterpillars simply sit on branches eating leaves and looking defenseless. What could something so fat and squishy-looking do to a large human being? Most of the time this thought is true, but some caterpillars have evolved a venom associated with their stiff spines.

When someone brushes against these caterpillars or tries to brush one off, these miniature hypodermic needles inject venom into tender skin which causes burning and welting. Some caterpillars announce their defense by bright colors and wickedly branched spines, but others may be like the proverbial 'wolf in sheep's clothing', masking those weapons under soft hairs. It is always good to be aware that Nature's creatures have many ways to protect themselves.

BUCK MOTH CATERPILLARS

(Hemileuca juno)

Buck Moth Caterpillar
Actual size.

Caterpillars with branching stiff spines all over the body are to be treated with respect, because all of them have some venom in the spines. Buck moths are the ones most likely to be encountered. One species, the palo verde buck moth, is most likely to be present in desert foothills areas in late fall and even winter. The caterpillars feed on foothills palo verde as well as the more ornamental blue palo verde.

Full-grown caterpillars are about an inch and a half long, pale, with a little marking on the body, and dark spines. A closely related species, the mesquite buck moth, has caterpillars that are blue-black with a series of dark-red spots on the back. This species hatches in mid-April as the mesquite starts to leaf out, feeds until May when it moves under debris on the ground and pupates. The adults emerge in the fall when they mate and lay eggs in bark crevices. The eggs overwinter and start the cycle over the next spring, thus avoiding the extreme conditions in the desert.

The Range Caterpillar, *Hemileuca oliviae,* another buck moth of the Southwest, is noted for its periodic population explosions in cattle-grazing areas, creating tremendous concern as the larvae compete with cattle for grasses. This moth also oversummers as a pupa, emerging as an adult in the fall to mate and lay eggs. Parasites seem to keep the populations in check most years.

PUSS MOTH CATERPILLARS

(Megalopyge bissesa)

Puss Moth Caterpillar
Actual size.

A hazard to children in rural schools in the oak zone is the very innocent-looking puss caterpillar. This one doesn't even look like a caterpillar. It's soft, furry, pale tan with small dark markings, and shaped much like a moth with its wings folded tentlike over its back. The largest of them are about an inch long. It feeds on the leaves of oak and several other trees and shrubs, reaching full size early in the fall, about when schools open. It is definitely a hazard to nature study in the lower grades.

The soft fur conceals short, erect hollow spines that are easily broken and deliver a very painful dose of venom. Handling the caterpillar by anyone is risky. The thin skin of small fingers makes young children particularly vulnerable. Medical attention is indicated if anyone gets stung.

The caterpillars make a hard oval cocoon, incorporating most of their fur, pupate within this shelter and emerge as moths the following summer, through a lid at one end that readily opens to allow the moth to escape this safe haven.

SCORPIONS

(Order Scorpionida)
(Centruroides exilicauda—Bark Scorpion)
(Hadrurus arizonensis—Giant Hairy Scorpion)
(Vaejovis spinigeris—Stripe-tailed or Devil Scorpion)

Almost everybody is familiar with scorpions as denizens of the desert, and the knowledge is reinforced by the use of these arthropods as ornaments of all kinds when embedded in plastic. Scorpions

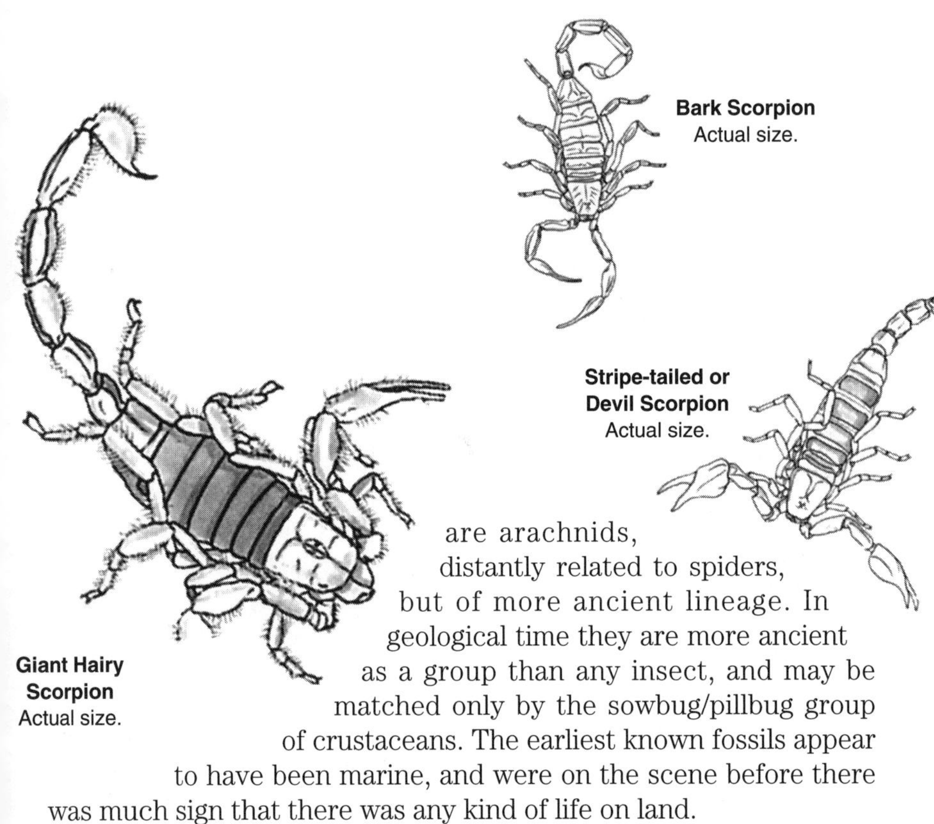

are arachnids, distantly related to spiders, but of more ancient lineage. In geological time they are more ancient as a group than any insect, and may be matched only by the sowbug/pillbug group of crustaceans. The earliest known fossils appear to have been marine, and were on the scene before there was much sign that there was any kind of life on land.

Everybody seems to know also that the working end of a scorpion is the tip of the tail, which has a sharp point and a reservoir of venom. The Southwest has many species of scorpions, from about two to more than four inches in length as adults. The largest is the giant hairy scorpion, scariest because of its size but possessing a rather mild venom. The bark scorpion, a slender straw-yellow species, possesses the most painful and long-lasting venom. Not to diminish its potency or the potential harm of an encounter, but the scare stories concerning death by scorpion sting are quite exaggerated. The bark scorpion is the only species I know that climbs vertical surfaces, which certainly adds to the horror of the homeowner's discovery in the bedroom or anywhere else in the house. The other common small scorpion is the striped-tail or ground scorpion.

Part of the lore of the Old West was the need to dump out one's boots in the morning to make sure no scorpions were in

them. This is a good idea in scorpion country, as is making sure that clothing is stored off the ground and also shaken out. All species are nocturnal, hiding away in dark places during the day. They seek out the most humid available location in their dry environment. This may be in holes in decaying logs, crevices in rocks, under rocks, or in similar places.

Scorpions are very sedentary, taking the same position every night for many nights in a row, waiting for potential prey to pass near enough to stimulate them to attack. The stinger is used to immobilize the prey while it is being held by the pinchers. In the attack the scorpion may be very athletic. They do move location occasionally, probably seeking a more favorable place to spend the day.

New houses in scorpion habitat produce the higher humidity from drying plaster and concrete, so are more likely to have a scorpion problem than houses a year or two old. Eliminating a scorpion population involves changing the habitat in and near the house. Woodpiles and rock piles are likely homes.

New construction in a neighborhood also leads to more scorpion activity as the scorpions are chased from their established territory. There are no nests of scorpions, but surprisingly the bark scorpions are gregarious, a peculiar trait for such a ferocious predator, and may be found clustered under an old piece of plywood. The bark scorpion seems to be becoming more urbanized, maybe because food sources such as crickets and roaches are becoming more available too.

One of the most effective ways of finding scorpion habitat is to search the area on a warm evening, especially during the dark of the moon, with a filtered ultraviolet light source, such as is used in light shows. All scorpions fluoresce a brilliant greenish-yellow. A small investment in a battery unit that operates a fluorescent tube, substituting a filtered UV tube, is a good one for anyone in scorpion habitat.

SPOTTED BLISTER BEETLE

(Epicauta pardalis)

Blister beetles get their name because they can cause blisters upon contact with the skin. The causative agent is *cantharidin,* a very active substance, which is contained in the blood of the beetles. If

the beetle is stressed in any way, such as by being pressed against the skin, it can exude blood from the joints in its legs and antennae. The cantharidin causes a mean-looking but very shallow blister, sometimes inches long and the width of the beetle.

The obvious function of the cantharidin is protection. Most predators avoid blister beetles, toads being an exception. Many blister beetles are brightly marked, as warning to the predator. The cantharidin is a very poisonous substance. Horses happen to be very sensitive to it. A very few dried beetles mixed in the hay may deliver a dose that causes severe colic and sometimes death. Other animals don't seem to be so sensitive.

The relationship between blister beetles and horses has not been widely known for more than about 20 years. It first gained publicity when hay from the Midwest killed horses in Florida and the poisoning was traced to the striped blister beetle that is prevalent there. The beetles had been harvested and dried with the alfalfa hay. Soon after, a similar problem in Arizona was tied to the spotted blister beetle, *Epicauta pardalis.* This species is about half an inch long, mostly black but with a scattering of gray spots made of hair, and soft-bodied.

There are many species of blister beetles in the desert Southwest (discussed under the iron cross beetle) but only the spotted blister beetle is likely to end up in alfalfa and get baled. Only one cutting of alfalfa is likely to be affected. Many species have the behavior of congregating soon after they reach the adult stage, a mechanism to bring the sexes together. Very often the congregation is guided by rather specific food plant preferences.

The spotted blister beetle is partial to legumes, and probably favors mesquite trees. Once an aggregation starts the beetles attract each other, until there may be a swarm of thousands. These can eat all the leaves off the tree and then move on, usually downwind. I have seen stripped trees in a straight line several hundred feet long. Smaller groups probably occur before the swarms start, and some of these build up in alfalfa. The range of the beetle is primarily in farmland above about 3000 feet, and the season of activity early summer. The larvae of this group of blister beetles feed on grasshopper eggs, thus a good crop of grasshoppers indicates a potentially good blister-beetle population the next year.

SPIDERS
(Order Araneae)

Most spiders, those eight-legged arthropods that can create fear and hysteria in movies and homes, really are gentle predators. They build webs to trap prey, and except for the dust-covered cobwebs termed unsightly by homeowners, spiders could be left alone to snare and eat those other bugs in the house. There are several which have developed a venom much more potent than the majority of spiders, and when people encounter these spiders, problems from bites may arise. It is not easy to get bitten by most spiders because of their shy nature. Many ailments blamed on spider bites are not justified. It is wise, though, to know these spiders and understand how to deal with them.

ARIZONA BROWN SPIDER
(Loxosceles arizonica)

Interest in brown spiders started in the Midwest and Southeast where an imported species, *Loxosceles reclusa*, the recluse, fiddleback or violin spider, is likely to live in closets and even inside clothing. We have a close equivalent here, *Loxosceles arizonica.*

Arizona Brown Spider
Actual size.

Even a spider expert has to look closely to tell the two apart. The color is, of course, brown, although the abdomen is gray. There is a violin-like marking on the front half of the body, which has given the Midwestern species the nickname of *violin spider.* The trouble is that brown is a very common color in spiders and many have a violin-like marking.

The body is fairly small, barely three eights of an inch long, moderately slender, but with a fatter abdomen in the female. The legs are long and not very hairy or spiny. The clincher for identification is the fact that brown spiders have three pairs of eyes, arranged in a crescent shape on the top of the head. Most spiders have four pairs of eyes in various arrangements. If you can count the eyes, you have gotten over your fear and distrust of spiders.

The other characteristic that distinguishes brown spiders is their web, a sheet of milky-white silk that does not have any kitchen leavings attached to it. This is a fastidious spider. The ordinary place to find the webs is under things on the desert, any place there is a space of an inch or so. The spiders do get into houses in areas of natural vegetation, perhaps especially new ones with more than normal moisture present. A typical way of bringing brown spiders into the house is to decorate with cholla skeletons, a favored treasure from the desert. If scorpions keep getting into the house, brown spiders may be found there as well.

In the Midwest the most frequent biting encounters occur when clothing is being put on. The spider is not aggressive, but responds to being squeezed. The effects of the bite are quite startling, starting with a blister and developing into an ulcer that refuses to heal. Medical treatment is certainly indicated. The physician will almost certainly want to see the spider if you suspect you were bitten by a brown spider. Finding the culprit, however, may be difficult. As a result treatment is usually based on symptoms.

BLACK WIDOW SPIDER

(Latrodectus hesperus)

Black Widow Spider
Actual size.

The adult female black widow is readily identified by the red hourglass marking on the underside of the shiny black (sometimes brown) abdomen. Because the widow hangs upside down in her web, this stop sign is always exposed. The abdomen, tear-drop shaped, is about half an inch in diameter. The front of the body that possesses the long, gangly legs is dwarfed by the abdomen. This is a strictly nocturnal species, hidden inside a silken funnel spun in a crevice at the base of the web. When this spider is surprised while exposed on her web, she rapidly retreats to that secure hideaway. The web itself can be several feet across, with guywires well beyond that, and is made of the strongest and stickiest silk of any of our spiders.

Most black widow webs are found in sheltered places out-of-doors, especially in sheds and on porches near the lights. Favored places will soon be reoccupied if the spider is removed. Much less conspicuous are the immature spiders, which have black-and-yellow stripes and spots on the abdomen and make correspondingly smaller webs. The immature pattern is carried into adulthood by the tiny males, one of which is almost always present on the fringes of the female's web. Immature spiderlings also start out with a orangish-white hourglass marking, turning red with age.

The sudden appearance of adult females in the spring marks the end of their hibernation. Later in the year it may also seem as if there has been a sudden outbreak. This is mainly because the immatures are so inconspicuous. Good rains during summer generate good populations of insects, and indirectly increase the growth rate of their major predators. There may be very few black widows in a dry year. The eggs are deposited in cream-colored silk cocoons in the web. Females are aggressive when they are guarding eggs, a definite change from their normally passive demeanor. The spiderlings that hatch from the cocoon soon disperse themselves and are seldom seen.

The bites of females must be taken seriously. Most people make it a practice to keep these spiders in check. I have found that a few trips around favored places with a fly swatter in the first warm evenings of springtime can reduce widow numbers drastically.

BITING FLIES

(Order Diptera)

The flies most people know are ones that probably just annoy by landing on your face. Biting flies, though, create pain because they have modified their mouthparts and feeding behaviors. The biting flies' mouthparts may be hypodermic needle-like, as the mosquito's, that take nice clean meals directly from one's capillaries. Little pain is felt but the ensuing itch drives a person crazy. Black flies and horse flies differ by ripping the flesh apart with sawblade mandibles, causing blood to pool up. These monsters then lap up the blood as a dog might water.

Sometimes only females are blood-feeders, an adaptation associated with egg production, while males feed on nectar. In some

Biters and Annoyers

Vertebrate blood is a particularly indigestible substance, but a variety of insects have "learned" to cope with it physiologically. Mosquitoes can be so numerous in salt marshes and on the tundra that they can drain a large animal of its blood. For us, the itching of the bites is usually the limit of our involvement. No-see-ums and chiggers can be very annoying in certain habitats. Avoiding areas where biters live is the only sure way to remain itchless. If that is not possible, use of a good repellent should help. Powdered sulfur, usually referred to as *flowers of sulfur,* is a good repellent for chiggers. Mosquitoes and biting flies are held at bay by several synthetic compounds easily obtained at pharmacies and outdoor-supply places.

groups, both males and females seek out the pleasure of people or their animals as dinner companions.

BLACK FLIES
(Family Simuliidae)

Part of the lore of the North Woods is the black flies that swarm around the exposed parts of the bodies of fly fishermen and bite them severely. The flies are small, about an eighth of an inch long, dark-bodied, and very persistent throughout the daylight hours during fly season. Resort areas in the Northeast and Canada essentially have to shut down when the season is on. They are a real plague. In addition to itching, the individual bites usually hemorrhage, leaving red spots in the skin. The larvae of black flies live in fast, clean, cold water, filtering out small organisms with special brushes on their mouth parts.

The southern coast of Brazil in the region of Santos is uninhabitable because of swarms of them that breed on rocky hillsides that run water all the time. This area has the highest rainfall in the continent. Pete Seeger, the American folk singer who sailed the Hudson River in New York on 'The Clearwater' encouraging people to clean up the polluted river, has helped bring back black flies to that region.

Of major concern to human beings is the buildup of black flies by irrigation and fishing project developments in Africa and Central America. The flies serve as vectors of an onchocerca worm that eventually causes river blindness.

The Southwest's black-fly problem is not of this magnitude. It happens that we do have black flies wherever there is clear flowing water, but most of these species consider human blood unfit to drink.

NO-SEE-UMS

(Family Ceratopogonidae)

No-see-ums were aptly named by Native Americans in the North Woods. Among the biting insects these are the only ones that are small enough that bites seem to appear without anything biting. They are tiny flies, and have been given appropriate, sometimes unprintable names wherever there are species that will accept human blood. Fortunately for us, most don't, preferring other vertebrates or even other insects.

Some parts of the west coast of Mexico are almost unbearable when the breeze comes from the no-see-um (known locally as *hequenes*) breeding areas. A stiff onshore breeze is a welcome relief. Ordinary screening and netting have mesh too coarse to keep them out, and fine-meshed screening keeps too much breeze out in hot weather. Repellents placed on the skin are about the only possible protection. The problem is worldwide. In England the beasts are called *punkies*.

This region's location away from the coast keeps us from most of the no-see-ums, but there are a few species that find a larval habitat here. Ordinarily, this is in rich canyons in the mountains. Any activity that stirs up the fallen leaves in such a place is likely to generate a pestilence of no-see-ums.

More annoying because it is likely to turn up in other places where people live is a species with larvae that thrive in disturbed clay soil. New subdivisions that have land levelling and ditch-digging going on are susceptible. Fortunately, the season of emergence of the adult flies is usually only a few weeks per year and the digging may have stopped by the next year. The problem is worsened by the tendency of the flies to seek a shady spot in which to rest. This can get them through screening and into the house.

STABLE, HORSE, DEER AND SNIPE FLIES
(Families Muscidae, Tabanidae and Rhagionidae)

The order Diptera is well provided with species that make a living on the blood of mammals, including man.

The stable fly, *Stomoxys calcitrans,* or biting house fly, is a denizen of barns and stables. The larvae breed in rather liquid manure. The adults look like house flies, but have a piercing set of mouthparts that delivers a red-hot bite. The flies seem to get hungriest when the humidity increases during the approach of a summer storm.

Most of the other biting flies live at higher elevations. An exception is an inch-long horse fly, *Tabanus punctifer,* black with white fur on the thorax. It can make a nuisance of itself along the main water courses. Its counterpart at higher elevations is the deer fly, *Chrysops facialis* that has brownish markings on its wings. Fortunately, they are easily swatted. If these flies do land, you immediately know because their method of feeding is to rip apart the skin and cause pooling of the blood that they lap up.

A very inoffensive-looking fly is the biting snipe fly, *Symphoromyia* spp. This creature lives in the rich meadows at high elevations. It is a very inept blood eater, and seems to have a hard time deciding where to bite. However, the bites are well worth avoiding, as they itch for days. Fortunately, the season of adult activity is brief, a week or two in early summer.

All of the higher elevation biters are associated with cienegas and other aquatic sites for the larval development. Removing oneself from an aquatic association will usually escape the problem of the bites.

CHIGGERS
(Order Acari)
(Family Trombiculidae)

Our tiniest itch generators are the chiggers, which are mites, Class Arachnida. They are red, but so small that only someone interested in studying them is likely to see them. This does not mean that they are not encountered. These are the creatures that crawl up the legs of unsuspecting people and work their way down into hair follicles before they start to feed. They inject a substance that probably has

the function of dissolving tissue. They feed for a period, then drop off. They leave behind an itch that persists for a week or more. Chiggers are prominent in tales of life in the Southeast, the Midwest, and Texas.

Many assume that chiggers aren't to be found west of Texas. Fortunately for most urbanites, this is true for the desert cities, no matter how much vegetation is generated by irrigation. They are present in the grassland and oak/juniper zones of the Southwest, and into the pine zone in some places. Their numbers are highest in late summer, and are increased by a wet season.

Chiggers are the immature stage of a group of mites that feeds on insects as adults. Protection from the bites can be gained with repellents applied to stockings, lower legs and arms. An old favorite repellent is flowers of sulfur, that has a smell that persists even through a washing. Mosquito repellents work for most people. Be cautious about sitting on the ground in chigger habitat, even with repellent on. High shoes and boots offer protection from infestation via the feet.

Cures for the itch they cause are legion, some of them probably effective. If you can tolerate smelling like a salad, I have found vinegar to be most effective at stopping the itch. Suffocation by fingernail polish is not worthwhile because the chigger is probably gone by the time you discover the misery of their digestive juices in your ankle.

EYE GNATS

(Hippelates spp.)
(Family Chloropidae)

The Desert Southwest is one of the most comfortable places in the country to be out-of-doors, but the nicest places in desert canyons are almost always infested with eye gnats during the summer. These are tiny flies with rasping mouthparts too small to be a bother. What is a bother is that they persist in trying to land in a person's eyes, ears and nose. Slapping them is difficult because of where they fly, and they never get discouraged.

They tend to avoid full sun during the hot part of the day, so when you sit in the shade, you are going into their environment. Fortunately, they are strictly diurnal. Before complete darkness they are gone for the night.

The immature stages live underground, on the roots of plants. The planting of gardens and lawns in urban areas usually changes the habitat enough that they are not a bad problem. This is not always the case, however. Date palm groves have them as a major pest.

Keeping gnats out of the ears is fairly easy with insect repellent. Keeping them out of the eyes is hard. Wearing a broad-brimmed hat tends to help. They walk around on the hat and don't get down under the shaded part to the face. They have been implicated in carrying the bacteria that cause human pink eye, but mostly they are a nuisance.

FLEAS

(Ctenocephalides felis)

Fleas in the Southwest normally are problematic to cats and dogs. If flea populations are not dealt with on pets, they can become a big problem biting people also. If you start to get red welts around your ankles, and your dog or cat is constantly scratching, you are in for a major battle.

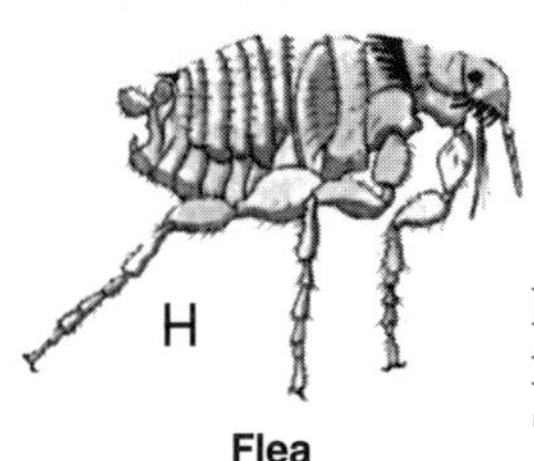

Flea
Fifteen times actual size.

Fleas are called *ectoparasites* because they search out a host, then live and feed on the outside of the host's body. Fleas are blood feeders, which explains the red, itchy welts people experience after a feeding by fleas. The adult fleas are small but visible to the naked eye. Their body is flattened from side to side, allowing them to move through animal hair easily. They have long, powerful hind legs that act as jumping organs. They have flattened spines arranged like a comb near their head, a device that keeps the flea from being easily dislodged by the host's scratching.

The immature stages of the flea include small white eggs that are laid in an area where the host, your dog or cat, probably spends a lot of time relaxing. The larval stage is a white, hairy maggot-like worm, about a fourth of an inch long. The larva feeds on various organic materials in the host's bed area, but requires the adult fecal pellets that contain digested blood to complete its metamorphosis. It spins a silken cocoon and changes into the adult flea.

In the Southwest, fleas normally don't become much of a problem unless it is a very wet year. In houses where humidity might be higher, one may have fleas most any time though. To deal with fleas, sanitation is most important. Treat your pet with appropriate medication. Clean the areas where your pet spends a lot of time laying down.

In hard-to-reach spaces, the use of diatomaceous earth, a natural substance made from the calcerous remains of marine animals called diatoms, may be necessary. The diatomaceous earth abrades the exoskeleton of the larval fleas, causing dehydration and eventual death. With diligence, this should control the problem. If the flea population has grown out of control because of neglect, one may have to resort to drastic actions using professionals to treat the home. New treatments include the use of juvenile hormones which can affect the development of the larvae, but these are still rather expensive.

Fearsome But Harmless

In this category is an assemblage of creatures that, because of their large size, appear as threats to our very existence. They may not be *completely* harmless. I was bitten in the Philippines by the largest centipede I have ever seen, over a foot long. Reference to my textbooks at the time provided me with the information that, "It is said that the bite of these arthropods is extremely painful." I agreed. A day of sweating and chills took care of the problem.

BOT FLIES
(Family Oestridae)

Real Southwestern hunters don't hunt jackrabbits because they have worms. Most of the worms referred to are huge grubs under the skin that the hunter really would like to avoid having move into him. Some rabbits do have the cysts of tapeworms in their muscles and these are to be avoided if you are in the habit of eating raw meat. Coyotes and other predators are the normal host for the tapeworms. But the big grubs under the skin are fly larvae and their choice of hosts is so specific that they never live in any other kind of animal. Most that are seen are nearly an inch long, blunt at the ends, white and covered with short brown spines.

They can be cut or squeezed out of the skin and will crawl around for some time after the excision. They are larvae of several species of *Cuterebra,* a large black fly with white markings and no functional mouthparts. The flies are sometimes seen sunning themselves in rabbit habitat, but they spend most of their time on hilltops, where boy meets girl. They fly amazingly well for an insect so obviously heavy-bodied, looking like oversized horse flies. The females must spend most of their time after mating in finding a suitable host animal. It is not true that they are limited to jackrabbits. Cottontails are infested also, but not as commonly.

Other species of *Cuterebra* concentrate on pack rats. One or two larvae under the skin really must inconvenience these animals. They are huge in comparison with the host. The bots keep contact with the outside through a hole in the host's skin, and leave through that hole when they are ready to pupate. The body shortens up and hardens to form a dark-brown puparium almost as hard as rock.

There are other groups of bots that live under the skin of cattle, in the sinuses of sheep, and the intestines of horses. The torsalo, a closely related bot found in the moist tropical highlands of Central and South America, lives under the skin of cattle and sometimes humans. The female solves the problem of spooking the host when she gets near by the simple device of capturing a mosquito or other biting fly and laying her eggs on that bloodsucker. The eggs are ready to hatch as soon as they encounter the warmth of a mammal, and drop onto their new host as their taxicab fly is taking a blood meal. The bot larva then enters the host through the feeding hole of the other fly. People who have gotten infested say that the larvae are very irritating, even painful. Cattle don't seem to mind, but develop oozing sores that make them very unattractive.

CENTIPEDES

(Scolopendra heros)

Centipede
About five and three quarters inches long, shown here one half actual size.

Centipedes are more a part of the desert scene than they are of other type of country. Everybody seems to know that the name means "hundred legs" and that no species has as many as a hundred. There is one pair per body segment, although the last pair are elongated and help hold prey, or startle the unwary handler. The commonest of the desert species is tan with weak darker bands, and about three inches long.

The giant of the group is six inches long, or even more, and has a green-black zone on both front and back end. Centipedes are active predators, entirely nocturnal, and spend their daylight hours under rocks or down other creatures' holes. When a centipede is exposed to the sunlight, its immediate response is to run for cover. This escape route may be in your direction, but that is coincident with shelter in your direction. The centipede is not attacking you.

A peculiarity of the group is that the first pair of legs has been modified into opposable fangs. Look closely at the head and you will note a pair of short, very stout legs outlining the head capsule. These legs end in sharp hollow claws. These are used to inject venom into the prey.

The mouthparts are not at all modified for predation. Unless you pick up a large one, you are in no danger of being bitten. The fangs lie on the underside of the head and do not show unless being used. For humans a bite is painful but not dangerous. One South American species is said to have a potentially lethal venom, although any venom certainly can be dangerous.

For some reason, the standard Old West story has been that all the centipede's legs have venom, and one walking across your bare chest will leave a trail of red welts. Not true at least from the venom standpoint, but if your skin is highly sensitive the little claws all falling at the same point might cause some irritation.

MILLIPEDES

(Orthoporus ornatus)

There are many kinds of millipedes, but to the Southwestern desert dweller millipedes are long, red-brown creatures with narrow paler bands, as much as six inches long. Millipedes have a cylindrical-shaped body in contrast to the flat centipede. Their name suggests a million legs but they fall far short of that number. They are the only arthropod to possess two pairs of walking legs on most of their segments, the front three or four segments having just one pair. They appear out of nowhere after a good rain in the summer, stay around for a while, leave some dead relatives on the surface, and disappear. They're not seen again until there's another good warm, soaking rain.

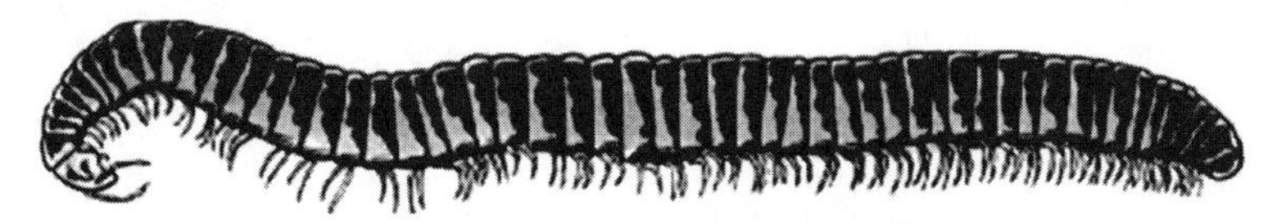

Desert Millipede
Actual size.

Those who have studied them have learned that they are plant eaters, feeding on both living and dead material. Their mouthparts are small and fitted for chewing, so their feeding isn't very conspicuous. When they go down into the soil as the season dries up, they follow the path of least resistance, using the passageways of deep-digging ants and probably any other openings. They have been found many feet down.

Their extremely hard body must keep them from being successfully attacked by stinging ants, although their numerous tiny legs would be subject to removal. Their main defense may very well be chemical. Every segment of the body, and there are a lot of them, has a pair of glands, one on each side. The glands exude a bad-smelling brown fluid when the animals are disturbed. So millipedes can take care of themselves.

Despite their armament and defenses, they do seem to be very vulnerable to desiccation. If they spend long on the surface during dry weather, they may not make it to a friendly hole when the time comes. Their only real enemy in the animal world is the larva of a beetle called *Zarrhipis.* This is a slender yellow-and-black-banded larva up to two inches long, with a vicious bite and obvious venom. It searches down millipedes, kills them with a bite and eats its way down inside from the front. The *Zarrhipis* larva wouldn't be very conspicuous were it not for the fact that it is luminescent. People on moonlight hikes have brought in the most specimens to the entomology department. The appetite of the larvae is such that we have never been successful in finding enough food to carry them through to the adult stage.

The only other millipedes likely to be seen are small, not much more than an inch long, and knobby-bodied. These turn up under things in patios. These do not have the defensive glands along the body.

TAILLESS WHIPSCORPION

(Paraphrynus spp.)

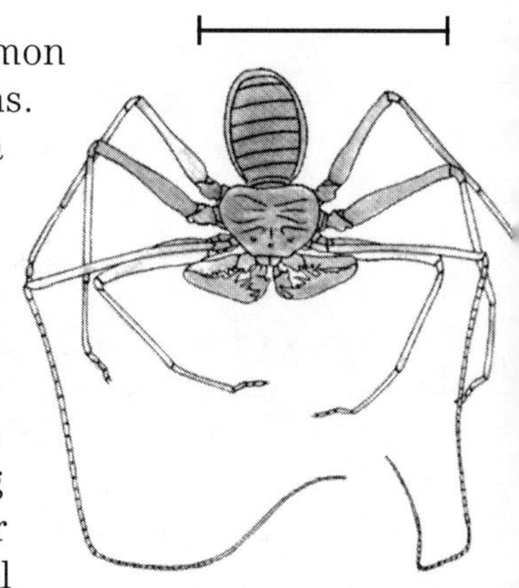

Tailless Whipscorpion
One half actual size.

In a wonderment of confusion in common names, these animals are *not* scorpions. They don't look much like a scorpion or a whipscorpion because they have no tails. Most people probably mistake them for an odd-looking, flattened spider. The front of the body is designed more like whipscorpions, with the peculiar heavily spined, grasping arms. The big feature, though, is a pair of very long and limber front legs, which are used for the sense of touch and probably for smell and taste. The color is a dark brown and the size up to a little more than an inch across the very flattened body. The tailless whipscorpion order Amblypygi is as ancient as the whipscorpions but is not closely related.

These animals are probably widely distributed at lower elevations, but are not often seen. One collector I know has found them on the ceiling of round-tailed ground-squirrel burrows. It's certainly easier to share a home with a good digger and stay out of the way, than to have to dig in caliche yourself. These strange creatures hunt at night and may wander into a garage where they are encountered by people. They are so bizarre that the discoverer has to learn more about them and brings it to our department.

Amblypygids are predators, moving slowly about with the long front legs exploring for what lies ahead. They have no venom. We really don't know much about them except that the same species ranges widely in North America and there are closely related species on other continents.

When they were exploring King Tut's tomb I noticed one on the wall in the film. To add to the confusion, the name given one genus is *Tarantula,* the same as the common name of our giant desert spiders. The name Tarantula really should have gone to one of the wolf spiders in Italy, which was blamed in the Middle Ages for causing *tarantulism,* a dancing frenzy, in the citizens of Taranto.

TARANTULAS

(Aphonopelma chalcodes)

Tarantulas are the gentle giants of our spider world. Several species are found in the desert areas of the Southwest, most having a body length of up to about three inches. All females are usually blond to brown, extremely hairy with heavy legs. Males are black with reddish hairs on the abdomen. Color patterns of tarantulas from around the world are much more picturesque than the southwest varieties.

Tarantula
Two thirds actual size.

All species are nocturnal animals, living in holes of their own construction. These holes are lined with silk, as one would expect from a spider, but are very inconspicuous. The tarantulas go to a great deal of trouble to scatter any soil that they remove well away from the entrance. Some other digging spiders are not so tidy. The holes are about a foot deep, and angle off to the side in a chamber where the tarantula spends most of its time. What isn't immediately obvious is that silk strands extend out from the nest entrance. These serve to notify the occupant when potential prey approaches the hole. Waiting for a meal is an almost full-time occupation during warm nights. If the tarantula is down in the hole and not waiting for prey, it places a thin layer of silk across the entrance.

Female tarantulas are heavier-bodied than males. Another sexual difference is that males have special hooks on the front legs just above the second joint. They use these to hold the female fangs during mating. The male must move beneath the female to about the middle of her abdomen to complete the mating process, using his modified pedipalps (the small leglike appendages at the very front of the spider) as a copulatory organ. After mating, the

male quickly backs out of his precarious position, releases the female fangs and runs.

Tarantulas have a mild venom and impressive-looking fangs, but are very unlikely to bite unless roughly handled. Their eyes are incredibly small. Four pairs are located on a small bump in the middle of the cephalothorax. They can't be of much use in forming an image, so the tarantula probably isn't aware you are there if you don't move or touch its strands of silk.

The main defense tarantulas have against consumption by predators is a patch of loose barbed hairs on the abdomen. When disturbed, they move their hind legs back and forth rapidly over this patch of hairs, dislodging them into the air where these hairs may lodge in the eyes or nose of the invader, causing irritation and cessation of the stalk. Sensitive people who come in contact with these hairs may break out in welts. Several people have gotten these hairs lodged in their corneas, so care should be taken if one keeps tarantulas as pets.

Do not become alarmed if your pet tarantula does develop a bald abdomen, because it will shed its exoskeleton and be blessed with a new complement of body hairs. Because they are so large, the process of molting and getting a new exoskeleton is spectacular. The process continues annually for as long as the animal lives, at least a decade for mature females. This event does cause great consternation to the owner, who assumes the tarantula is sick or dying, because it usually ceases to eat and becomes quite lethargic. One day the spider may be seen on its back, but don't give up hope. Return within a day or two and you should find what appears to be two tarantulas where there once was one. Look closely and you will find one is simply an empty shell while the other is a beautiful healthy spider. The shell is quite interesting for it looks as if it simply popped its top (the cephalothorax in this case) and pulled itself out.

Tarantulas seen wandering on the desert at night in the summer are all males out looking for mates. Sometimes they are quite abundant. The pet trade handles several Mexican and Central American species that are even larger than ours. Tarantulas make quiet pets. One thing that must be avoided is dropping them. They are heavy enough that even a short fall can break the exoskeleton and suturing up a tarantula does not solve this fatal flaw.

WHIPSCORPION

(Mastigoproctus giganteus)

Whipscorpions or vinegaroons are found in the southeastern oak zone of Arizona eastward across the southern U.S. to Florida. With a substantial but flat body two inches long, huge spined, armlike pedipalps out front, front legs that are thin and elongate to act as feelers and a long whip-like tail on the behind, they look definitely Ante-Diluvian. They have no venom. They are arachnids, in a group of their own equivalent to the spiders, ticks and scorpions, and do have a long fossil record.

They are found in damp places, under logs and stones. They seem not to be able to dig very well, so they will be found in pre-existing holes. They move very deliberately when not pushed, but make surprisingly fast dashes if disturbed. They are predators, active at night, feeding on insects and other small animals in the right size range. Young ones have some of the appendages a rather bright red-brown, in contrast with the rest of the brown body color.

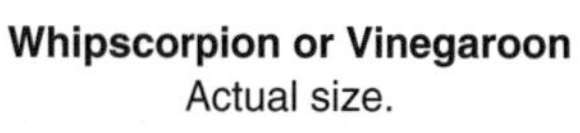

Whipscorpion or Vinegaroon
Actual size.

Their defense lies in the rear, associated with the whip-like tail. Acetic acid, vinegar to salad lovers, is produced from a gland in the rear, squirted out and spread around by the tail to inundate its rear, always a vulnerable spot for predator attack. Thus we get the common name *vinegaroon.*

WINDSCORPIONS

(Order Solpugida)

Windscorpions, sometimes also called *sun spiders* or *solpugids,* the last from the scientific name of the group, are pale-tan arachnids. The body is as much as an inch and a half long, with a pair of heavy pinchers dominating the front end. Being arachnids, spider relatives, they

have four pairs of long legs, and a pair of long appendages (pedipalps) in front that are used in touching and probably smelling. Instead of claws these appendages also have a terminal, eversible disc that aids in climbing.

Windscorpion
Actual size.

Windscorpions get the name from their speed and general resemblance to scorpions. They can run faster than any other arthropod one is likely to see, and they are extremely active predators. They subdue their prey with their pinchers, which lack poison glands. They would be capable of delivering a fairly strong pinch if caught, an unlikely eventuality because of their speed. They have pads on their feet that permit them to climb amazingly well. For this reason they may turn up in houses, and be seen climbing the walls. They seem to require a desert environment, and rarely survive urbanization.

A myth associated with these creatures leads to the Spanish name *mata venado,* or deer killer. Their grotesque appearance leads people to think them poisonous (which they aren't) and with their speed it was thought they could actually run down a deer, bite it and kill the deer with this supposed venom.

There's Millions of 'em

Some insects form such massive aggregations that the group seems to take on its own existence. We get calls from concerned citizens that there is a big blob moving down the road, changing shape as it moves along. What is it and how do we get rid of it? Some of these groupings are the result of a large population facing starvation when its food has been eaten or dried up. Others are the result of aggregation for overwintering or oversummering. Lady beetles on mountain tops are a familiar sight, and a real annoyance to astronomers. They tend to leave orange excrement wherever they walk.

FALSE CHINCH BUGS

(Nysius raphanus)

When somebody calls into the department of entomology describing a weird brown mass that keeps changing shape and moves slowly along, the immediate diagnosis is a migration of false chinch bugs. The phenomenon occurs mainly in disturbed lands next to agriculture in the lower deserts of the Southwest. The individual bugs in the mass are less than a quarter inch long when fully grown and winged, the nymphs smaller. When stepped on, all stages can emit the familiar stink-bug aroma, their protection from being eaten by some predators.

The plant largely responsible for the large numbers of false chinch bugs is the London rocket, a European weed in the mustard family that produces a spike of yellow blossoms in winter and early spring. The timing of the swarms of bugs is usually tied to the maturation of the plants, at which time the seeds mature and the plants dry up rapidly. The bugs may move onto crop plants at this time, but have never posed a serious threat.

As is usually the case, doing nothing about false chinch bugs produces about the same results as trying to kill them. They are with us all of the time, but the excessive numbers in late spring probably get cut back severely when they fail to find shade as we begin to celebrate summer in earnest in June. At that time they spend the daylight hours down in cracks in the soil, in the shade of plants or possibly on your porch, bringing dramatic newspaper articles about bug invasions.

These are insects with piercing, sucking mouthparts. This long flexible beak is able to penetrate soft plant tissue to supply this bug with a liquid diet. They got their name from a close resemblance to chinch bugs, which are serious pests of grain crops in the Midwest and Northeast. Chinch bugs are strictly associated with grasses, moving from small grains to corn as the season progresses. Barricades to their movement were the preferred method of control at one time, now largely supplanted by insecticidal control. One odd species of chinch bugs occurs in Arizona, but it is confined to St. Augustine grass in a case of overspecialization.

FLYING ANTS AND TERMITES

(Orders Hymenoptera and Isoptera)

Even though ants and termites are not closely related, the appearance of winged individuals from established colonies creates a common phenomenon that happens every year, usually tied to certain types of rain episodes or rises in humidity. These are the males and females of the species, which are engaged in a mating flight and have the potential of starting new colonies.

In the ants the females are capable of keeping the sperm alive after the mating flight for many years, producing fertilized or unfertilized eggs depending on what types of individuals (female or male) are needed for the colony. The oldest colony of ants I know about personally is a honey-ant colony in our driveway that has been there for 25 years. It looks the same as it did when we moved in. The mated female, termed a *queen,* lives a long life. The wingless worker ants probably live less than one season.

Termites have kings as well as queens. The mating flights in termites are more like mixers than mating flights. When the flight is over, males follow females, and a couple will attempt to start a new colony. The makeup of termite colonies is very complex, with male and female workers assuming many roles, plus having the ability to change their roles.

In both ants and termites the wings are used only once, being shed immediately after the flight. Queen ants are limber enough that they can twist the head back to the base of the wings and chew them off. Termites have a better design in that there's a weak line across the base of the wings. A quick flip and they're off. Who needs wings if you are going to live underground the rest of your life? The flight muscles are also a good source of energy to keep the queens going until the first brood of workers matures and begins taking care of the home front.

Every species responds to some particular combination of environmental cues, usually associated with warmth and moisture in the soil. Some fly during a summer rain, before, after or the next morning or evening. The critical thing is that winged forms of more than one colony be in the air at the same time. Events in the colonies underground have produced the winged forms. They may attempt to leave the colony ahead of time, but the workers keep them from doing it.

Such an abundance of animal matter in the habitat provides a reliable dinner for many insectivorous animals. Summer nesting desert birds are started on their nesting pattern by the food they eat. During a normal year this signal comes at the beginning of the summer rains for many species. We amuse ourselves thinking that this happens on San Juan's Day in late June, but mid-July is a more realistic expectation. Birds that have nested on this signal are set up to have a season of insect abundance in the form of flying ants and termites just when their nestlings need food most.

During one of our desert study years there was a really wet early June. The birds got turned on to nesting early, but the cues to the flying ants and termites were wrong. They were in the nests but did not fly. One entire brood of nestlings starved to death. Even some seed-eating species feed their young on insects, a concentrated protein diet.

If even one percent of the new colonies survived, there would be solid ants and termites everywhere in a couple of years. Survival past establishment must be a rare event. Seed-eating ants such as the harvester ants are closely attuned to the seed production of the site. A good grass stand generates a large number of colonies. Because the length of time a colony lives depends on the life span of the queen, the number of colonies is probably more stable than the stand of grass. Unusually few colonies mean that the grass stand is increasing, unusually many the stand is declining. The ants may interfere seriously with the reproduction of the grass, and have been the object of chemical-control measures in some rangelands. Establishing a new stand of grass is difficult if seed-eaters are well entrenched.

These soil dwellers are quite important, though, to the enrichment, aeration and workability of the soil throughout the world. They are as good as earthworms in soil turnover in the Northwest. When one thinks about the desert Southwest where earthworms for all practical purposes do not exist, we must look to the ants and termites to make the soil usable for plant growth. The seed-eating ants help disperse many of our native plants. These ants may not move rubber-tree plants but they have a huge impact on our ecosystem just the same.

SPRINGTAILS

(Order Collembola)

The springtails create much indecision amongst the entomological name-callers these days. Are they insects or just a close relative? Because of some technicalities about their mouthparts and other strange body features, they have been set apart from what we call insects. They are called *hexapods* because they do have six legs.

Springtails are minute, usually black, and have the ability to hop. Many times they are mistaken for fleas because of this leaping ability. They don't have jumping legs like fleas or grasshoppers, rather they use a tail-like appendage to hop. This appendage, called a *furcula,* folds underneath the body, hooks under a catch to build up energy, and then is released, causing the springtail to hop.

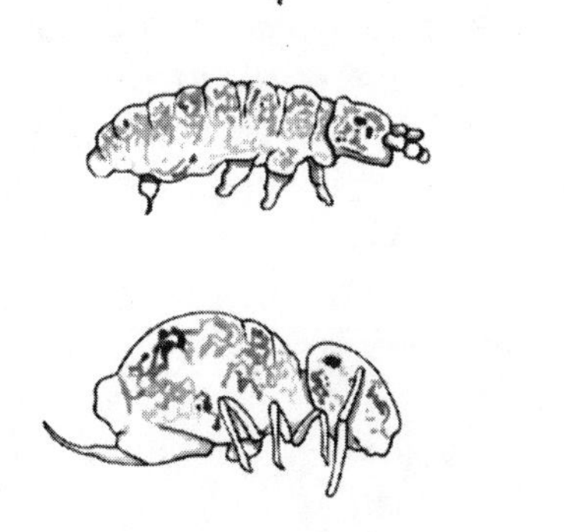

Springtails
Shown ten times actual size, each specimen would measure approximately two millimeters in length.

Springtails live in the soil, leaf litter or under tree bark. They feed on decaying plant material, bacteria, fungi, pollen or arthropod remains. Occasionally these hexapods have become problems in greenhouses or mushroom cellars, but generally they are innocuous. They may come indoors somehow, and many times people will find them in the bathtub or sink where moisture abounds. Because they are not biters or disease carriers, the easiest way to remove them is to wash them down the drain. If they are on the carpet, using a vacuum cleaner is probably the best strategy. When enormous numbers keep appearing as in Mesa, Arizona, a few summers ago, one need only wait for some cold weather and the population will recede. Because springtails are important soil-dwellers, the use of chemical warfare against them is unwarranted.

WHITE-LINED SPHINX CATERPILLARS

(Hyles lineata)

During the fall in a year of good summer rainfall, foothills and desert grassland areas of the Southwest can generate huge numbers of three-inch-long, hairless caterpillars with a narrow horn on the rear end of the body. They come in variegated colors of green with some black dorsal spots or solid-black stripes. These are the larvae of the white-lined sphinx moth. When they start their wanderings they can be so numerous as to grease a highway with their bodies or drive a pool owner to distraction.

Fortunately for the gardener, they don't seem to be very much interested in food. The preferred food plants are various wild relatives of the garden four o'clock. In an outbreak year the caterpillars eat all of them and look for more, transferring their attention to many desert annuals. They don't start to wander until they are fully grown. The trip ends when the individual caterpillars select a spot and dig down a couple of inches, where they make an earthen cell and pupate.

In one of our desert ecology projects we happened to have a plot where the caterpillars dug in. There were so many of them that the soil looked tilled. We anticipated hearing the roar of wings when the adult moths emerged, not being sure exactly when that would happen. It never happened. Quite obviously, the survival rate is not too high. The moths do turn up at lights and at blossoms in considerable numbers, but we have never seen numbers even close to the number of caterpillars. Natural control factors keep the population in check most of the time. Recent pollination studies indicate these moths are important to some night-blooming *Cereus* cactus species.

BANDED WOOLYBEARS

(Estigmene acrea)

Woolybears are blond to red-haired caterpillars that build up populations late in the season on cotton and wander around after the cotton matures. They are an inch and a half long and quite thick because of the dense erect hairs all over the body. The name is very descriptive. The official name of the Entomological Society of

America is *salt marsh caterpillar,* a name based on this species' preference for salt-marsh plants along the Eastern U.S. Coast. Quite obviously, they do not depend on marshes, salt or sweet, in the Desert Southwest.

The adult moths are rather striking, females having white wings with small black spots and an orange abdomen with a row of black spots down the middle. The males are similar but have orange hind wings also. They are attracted to lights at night, and may become numerous on lighted buildings in agricultural areas. There are several generations per year, the fall one usually the largest. The moth lays eggs in masses, covered with scales from her body. The caterpillars feed on leaves, and have typical caterpillar appetites.

Because the big populations occur mainly late in the season, insecticidal control is not usually practiced on cotton. The migration of the caterpillars into adjacent housing developments brings a hue and cry from homeowners. Barriers were once used against these migrations, the choice being for erect bands of aluminum foil six inches high. Hiring neighborhood kids to collect these cute furry worms is a great way to develop rapport and good will.

The moths, and probably the caterpillars, have their own chemical protection. There is no noticeable odor, but animals that try to eat adults quickly spit them back up. This is just another example of an insect that really stands out in its normal environment being chemically protected. The smart predator quickly learns to avoid that kind and kills very few learning about them.

Flashy in the Desert

Some insects and other arthropods are so large and brightly colored that they form a conspicuous part of the fauna. This is true of the southwest desert. Many of the animals are nocturnal or concealingly colored. The few that aren t really stand out. Incidentally, the brightly colored ones would be poor choices if you are ever faced with starvation on the desert. Dr. Murray Blum, insect physiologist, has a beautiful slide of an opossum reacting to a horse lubber grasshopper. He calls it his pukin possum slide. It s a crowd-stopper.

BLACK WITCH
(Erebus odora)

This is not a very abundant moth in the Southwest, but it is one that creates interest when it appears. It is large and broad-winged, brown with a variegated darker-brown pattern, and four or more inches in wingspan. The moths are most often encountered near lights on the sides of buildings. They usually have very tattered wings, an indication of old age or a long flight.

The moth is much more abundant in tropical areas, and may very well be flying in from lowland Sonora, Mexico to the U.S. deserts. We do find individuals with very fresh wings, however, an indication that they sometimes breed here. The caterpillars feed on mesquite tress, which we have in abundance, and probably on other leguminous trees. The moths generally appear in late summer.

CICADA KILLER

(Sphecius grandis)

Flashy, heavy-bodied, brown-and-yellow wasps an inch and a half long are cicada killer wasps. The females are great diggers, and are most often conspicuous because they favor sandy soil under sidewalks. This obviously cannot be their native habitat, but it is the one they most often choose in a city. The homeowner gets annoyed at the sand piles on top of the lawn and concerned that the walk may cave in if the use is heavy.

Those who are annoyed at the piercing calls of male cicadas can get satisfaction hearing the sudden drop in pitch and cessation that goes with the female cicada killer finding him. She quickly delivers a sting into the cicada s central nervous system and carries him back to the cell she has made in that tunnel under the sidewalk. There she lays an egg on him, seals off the cell and heads out to repeat the process.

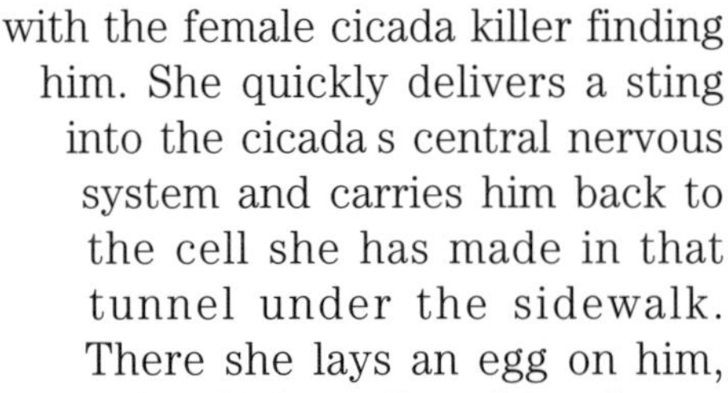

Cicada Killer with Cicada
Actual size.

All wasps except the social paper wasps and yellow jackets provision for their young similarly with variations on a theme. It is just the large size of wasp, nest and prey that make cicada killers so noticeable. The Southwest has two species, *Sphecius grandis* and *S. convallis,* one slightly larger and darker than the other. The appearance of the adults and the start of new burrows is correlated with the cicada season.

Large numbers of these giant wasps may be seen flying low to the ground. People mistake this cruising for attacking and may panic. Remember that these wasps are only searching, either for nest sites or their prey, not people. Stinging incidents are rare, for here is another shy giant. I remember growing up in Ohio, a neighbor in total panic used up four fire extinguishers to dispose of six cicada killers that were flying about in her yard, not knowing how harmless they really were.

CREOSOTE LAC INSECT

(Tachardiella larrae)

A few insects are included because of their use in indigenous artistry. Such are the lac insects that live on creosote bush. These are related to the lac insects from which shellac is produced in the Mediterranean region, and produce the same lac material. A group of lac insects on a creosote bush looks more like a heavy crust of reddish material, with drops of pale goo coming out of it. The mass is the product of a group of sucking insects, much like mealybugs or cochineal scale insects. Each has glands on its body that secrete lac instead of the wax that the other species mentioned produce. The lac is liquid at first, and soon forms a solid covering over the individual and coalesces with the lac from neighbors. The pale material coming out of the mass marks the hole that the insect has to the outside, and through which it passes the honeydew that it excretes.

Insects that suck the sap of plants all have two problems. One is excess water taken into the digestive tract. Most are able to pass through the excess. The other is that the food taken in is much higher in sugar than in amino acids and other nutrients. The excess sugar has to be gotten rid of. The result is honeydew, sugar water that often dries to transparent globs of sugar.

The lac insect on creosote bush is not particularly common, so the lac used to seal baskets for waterproofing must take a long time to accumulate. Another bush of warm canyons in the Sonoran Desert, *Coursetia*, has another species of lac insect, *T. coursetiae,* known as Goma Sonora in Mexico where it is made into a tea and used as a medicinal aid. The bush is an inconspicuous legume, notable for lacking spines or thorns.

CREOSOTE WOOLLY GALL

(Asphondylia auripila)

The most conspicuous insect-caused gall in the desert is the woolly gall on creosote bushes. This is a one inch spherical growth on the plant that is bright red-brown, definitely woolly, and

Creosote Wooly Gall
Actual size.

long-lasting. It looks more like some plant disease than the result of an insect s activity.

The insect that causes the plant to produce the gall is a tiny dark fly, not unlike the fungus gnats that sometimes become a nuisance in houses. The female fly deposits an egg in a terminal bud on the plant, apparently introducing a specific substance at the time. The egg hatches into a tiny red grub. As it feeds the plant grows the gall around it. Whether the growth is strictly in response to the substance left by the female or not is a difficult question to answer. The feeding of the larva is probably required to keep the gall growing. The larva may be adding more substances. Several different gall flies attack creosote bushes, each causing the plant to produce a particular kind of gall, of which the woolly gall is the largest by far. A graduate student at the University of Arizona discovered about 12 different leaf galls on creosote. Living inside a gall is a neat way to avoid the many defensive chemicals that protect creosote from the herbivores of the desert.

FAIRY SHRIMP/TADPOLE SHRIMP

(Brachinecta lindahli)
(Triops longicaudatus)

One of the most amazing phenomena of a desert waterhole is the sudden appearance of great numbers of animals, up to an inch long, with many legs. These are fairy shrimp and tadpole shrimp, both crustaceans but not very closely related to shrimp. The group to which they belong antedates that shrimp by a hundred-million years or so in the fossil record.

Fairy Shrimp
Actual size; shown swimming.

Fairy shrimp are slender and almost transparent, with a bulge behind the middle that marks the egg sac in the female. They swim upside down. Males have grossly enlarged second antennae used for clasping females. A species in the Southwest, *Brachinecta packardi,* has even added peg-like teeth to the antennae for better holding power.

Tadpole shrimp look vaguely like tadpoles, even more like the extinct trilobite group. They are brown, with a racquet-shaped

forebody and a slender abdomen bearing a pair of tails at the end.

The water in which these animals is found may be clear, but more often is a mud-bottomed cattle tank and thoroughly churned up by the shrimp. These animals feed on microorganisms in the mud, and complete their development in a few weeks. They are very well adapted to life in the desert. By the time the tank is dry they have mated and produced eggs. The eggs are very resistant to drying and heating, and will be there the next time the tank fills with water. It is assumed that the resistant eggs get moved from pond to pond in the mud picked up on the feet of birds.

Tadpole Shrimp
Actual size.

GIANT MESQUITE BUG

(Thasus gigas)

One of the flashiest insects on the continent is the giant mesquite bug, which ranges into southern Arizona from an extensive range in Mexico. The adult is two inches long, brown with yellow markings on the forewings. The legs are quickly noticed because of the black-and-red banding. The nymphs are banded with red and white. The male has grotesquely enlarged, strongly bowed and toothed hind femora. Probably the most characteristic feature of these giants is the flat, round disk towards the end of the antenna, a specialized sensory spot. The giant Mesquite bug feeds by sucking up plant juices and is extremely fussy about what plant it will use as food. It will accept all species of mesquite, but nothing else.

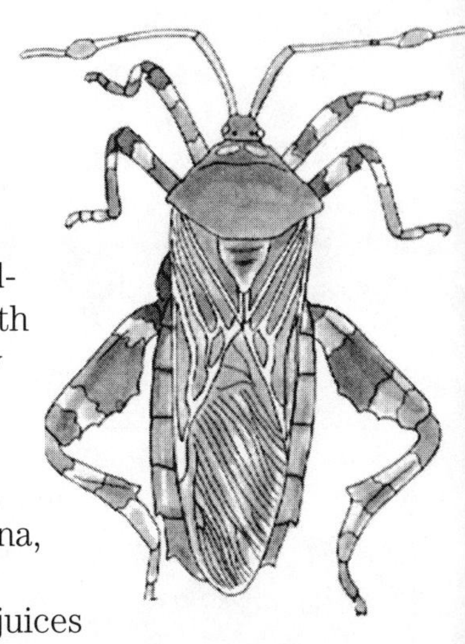

Adult Giant Mesquite Bug
Actual size.

Giant Mesquite Bug Nymph
Actual size.

The bugs tend to form colonies on the same trees year after year, but these colonies do not seem to be very extensive. There is one generation per year, the winter being passed in the egg stage. The eggs themselves are distinctive, like a row of barrels glued to each other end-to-end, on a twig or other surface. The nymphs are gregarious for a while after they hatch, usually in early April as the mesquite begins life anew. The larger nymphs do tend to aggregate and move up and down a tree depending on temperatures. All stages have well-developed scent glands, the nymphs in the middle of the top of the abdomen, the adults on the sides of the thorax just above the base of the middle and hind legs. These are the positions for all Hemiptera that have scent glands. Adults can accurately squirt the scent fluid, a fairly sweet smell, for several inches.

Surprisingly, many people have never seen these insects. They are abundant enough, but do live in trees and never reach large enough numbers to have any effect on the tree. They feed on sap. An adult feeding from a small twig looks as if it is out of scale with its food supply. Green pods may be used, but apparently for the sap and not for the developing seeds.

A related bug, *Acanthocephala granulosa,* is brown with some orange markings on its legs and the tips of the antennae. It is not as large as *Thasus* and certainly not as conspicuous. Males have the hind femora enlarged and bowed also. This species concentrates its efforts on developing seeds, in the same way that the leaf-footed plant bugs do. However, it is more partial to the pods of Yucca and Agave, and may become abundant on those plants. The feeding causes spots and irregularities on the pods.

HORSE LUBBER GRASSHOPPER

(Taeniopoda eques)

Perhaps the least believable insect around is the horse lubber grasshopper, a heavy-bodied black species with yellow to orange stripes and antennae, green-veined fore wings hiding pink hind

wings in the adult. They are a common sight on roads in southeastern Arizona every fall, when they and the spotted brown plains lubber grasshopper come to grief with our traffic and have a cannibal feast on their brethren. The horse lubber is a very noxious-tasting insect, turning back even the feeding attention of toads, notorious gourmands. It is surprising that the horse lubber and the plains lubber can tolerate it, but this is an observable fact. As has been noted in other conspicuous insects, the bright colors are a warning to would-be predators, although praying mantids and centipedes must be color-blind and have no taste because they are seen feeding on these grasshoppers.

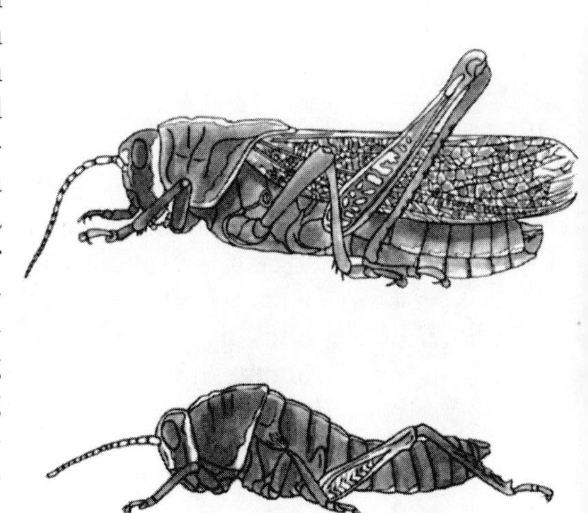

Horse Lubber Grasshopper
Adult, top; nymph below; shown actual size.

The grasshoppers that make it to the road show no inclination to fly, but the males can fly. They move from tree to tree stridulating by rubbing hind wings on front wings to excite females. The hind wings are specially modified to serve as sounding boards. The females put their eggs into a pod in the soil after mating, and the young hoppers get onto mesquite trees the following year. They are gregarious at first, then scatter. There is only one generation a year. They feed on a wide variety of plants, thus limiting their impact within the habitat for a population of giants like these could have a major impact if they fed on just one or a few kinds of plants.

The horse lubber is the far-northern representative of a group of grasshoppers that is conspicuous over wide areas in the New World Tropics. Some of the other species remain gregarious throughout life. All, including the horse lubber, emit a noxious froth from the spiracles on the side of the thorax, hissing as they do it. They just don't take any chances with predators.

IRON CROSS BLISTER BEETLE

(Tegrodera aloga)

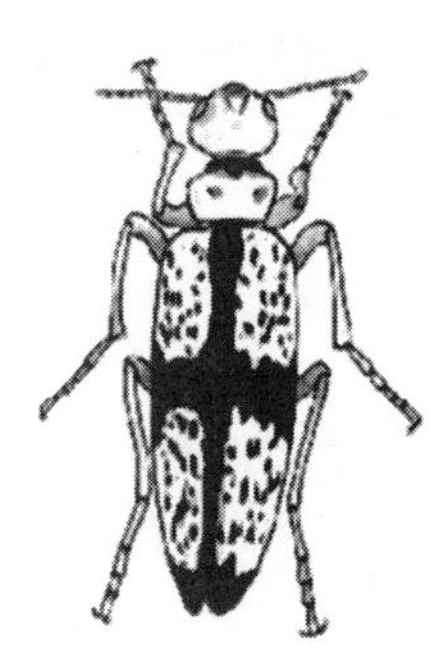

Iron Cross Blister Beetle
Actual size.

This is a species of blister beetle (see spotted blister beetle for discussion of cantharidin and blistering). It is a spectacular one, up to an inch and a half long, with the elytra variegated black on a yellow background with the head and thorax bright orange. It is an early spring member of the fauna of the lower desert, feeding primarily on the blue blossoms of *Nama hispidum,* or on *Eriastrum;* an inconspicuous annual. It can be abundant enough that it gets picked up, and brought to school for show and tell. Despite the fact that it is a blister beetle, we haven't heard of anyone being affected by its blood. It apparently does not aggregate very strongly. A spectacular behavior displayed by this beetle is seen when it is walking about the desert floor with its elytra (hard forewings) held up and its bright red-orange abdomen displayed for all to see. It reminded me of a spider wasp in hunting mode.

At least one other species, *Lytta magister,* can be abundant also, this one tending to form swarms, which may be noticeable to a passing motorist. It can be even larger, up to two inches, bright orange in front and black behind. It also lives on the lower desert in the spring.

Both of these species belong to one of two divisions of the blister-beetle family, Meloidae, that depend on the solitary bees as a source of food for their larvae. The females lay huge numbers of eggs in a hole in the soil. The hatchlings have to find the nest of a solitary bee, get in and kill the bee's egg or larva, then consume the contents of the cell. Such huge beetles as these must have to consume most of the cells in a nest to complete their development. It must be a rare event for the bees to build up to numbers great enough to produce a swarm of thousands of large beetles. Such swarms are never seen in the same area in consecutive years.

The other division that depends on bees uses the device of laying eggs on or near blossoms, the hatchlings being specially modified to attach themselves to visiting insects. If they happen to get onto a solitary bee building a nest, they have won. Even with this advantage the success rate must be tiny. A plant that is typically involved with this cycle of nature is the thistle *Cirsium.* Check for yellow egg masses underneath a developing bud, then check an open flower and you will probably be rewarded by finding an adult beetle feeding in the nectaries. Examine the flower more closely and you might even find the triungulin, a special name for the mobile larva of this beetle that is waiting for its limo ride to dinner.

The genus *Epicauta,* of which the spotted blister beetle described under *Poisonous and Venomous Creatures* earlier in the book is an example, differs from most other blister beetles in having the color made up mostly of a solid coating of pressed-down hair. All species in this group of blister beetles depend on the eggs of grasshoppers. Females lay eggs in the soil and the hatchlings have to find grasshopper egg pods. If preferred adult food plants are close to areas favored by grasshoppers for egg-laying, *Epicauta* species are helping to keep grasshoppers in check. Gentle slopes are preferred for oviposition by grasshoppers, so some encouragement of blister-beetle food plants there should help in preventing grasshopper outbreaks.

MESQUITE TWIG GIRDLER

(Oncideres rhodosticta)

Mesquite Twig Girdler
Actual size.

When the mesquite twig girdler hits, the mesquite haters are jubilant. The girdler is a rather attractive longhorn beetle an inch long, brown with a peppered paler pattern of scales. The beetles emerge in early summer, and very soon start providing for the next generation.

The female engages in the activity of girdling, using the mandibles to cut a deep channel around a half-inch branch. Then she goes out beyond the cut, bites several holes and inserts an egg in each. The leaves on the branch soon die and bleach, making a very impressive show for a little activity. The larvae that hatch from the eggs bore first into the cambium, then deeper into the wood.

Most times these elaborate plans for the babies don't produce great results. The smell of dead mesquite branch brings in other wood-boring beetles, which also lay eggs in the branch. Their larvae almost always manage to get to the food first, beating out the rightful owners. As a result a region that has had acres or even square miles of trees pruned by branch girdlers rarely has much sign of them the following year.

Mesquite is a tough plant; all the beetles do is provide a little pruning, actually helping those former distraught bee keepers and leaving the mesquite haters with lusher trees to complain about next year.

PINACATE BEETLES

(Eleodes obscurus sulcipennis)

Pinacate Beetle
Actual size.

Pinacate beetles are very durable-looking black beetles that spend a lot of time standing on their heads. There are many species of them, and could have been called Model T Ford beetles in another era because they come in any color you want as long as you want black. Some are more than an inch and a half long. Occasionally an odd beetle shows up with what appears to be a tail, as in the illustration. These individuals are males of several other species. The tail is an elongation of the elytra (front wings) a reverse-horn development probably evolved for impressing the ladies.

They are very characteristic of the desert, being found almost everywhere and throughout the year. They are active both day and night, but may spend the hottest part of the day down rodent burrows and under rocks. One of the sports of beetle collectors is to lay out a trail of oatmeal and see which pinacate beetles get to the bait.

The larvae are like the mealworms that are sold for bait and bird food in pet stores. They feed on dead plant material, usually where there is a little more moisture than in the surrounding soil. The adult beetles themselves make undemanding pets, thriving on a little fish food and a water supply. A bit more moisture, food and soil will support the larvae underground.

The head-standing habit is a way of displaying the defensive glands in the tip of the abdomen prominently for an approaching predator. The scent and taste protect them from almost everything. The grasshopper mouse has developed a way around this defense. It just grabs the beetle, pokes its behind down into the sand and eats it from the front end. Half beetles turn up regularly if you look closely. These creatures are so hard that they last years in the elements, just getting whiter as they age.

Native American mythology has wonderful stories that explain many of the strange behaviors of insects, the Pinacate beetle being no exception. The Cochiti of the Southwest associated this head-standing behavior with a creation story. This story noted that long ago *Eleodes* was assigned to place the stars in the sky. Unfortunately, the beetle became careless and dropped the stars that then scattered to form the Milky Way. Being so ashamed for what he had done, the beetle even today hides his face in the dirt when anyone approaches.

Another group of hard black beetles, *false pinacate beetles* or *darkling beetles* as they are called, have no repugnatory glands. These beetles too, walk about the desert, but when confronted by predators are probably just gobbled up. Many of the behaviors and food preferences are the same as pinacates, as all these beetles fill the desert scavenger roles.

Another darkling beetle in the desert, known as the *ghost beetle, Asbolus verrucosus,* has evolved a unique survival adaptation. When the humidity drops very low, it oozes a white wax from pores in the exoskeleton. The familiar black beetle now enshrouded in a waxy sheet can retain more water in this stressful time as it walks around its habitat as "Casper the friendly ghost beetle". Relatives in the Namib Desert of western Africa have developed colored waxes that blend with the various sands they inhabit.

One of our treasures is a batch of darkling beetles from a bottle found on the Camino del Diablo. What must have happened was that the first beetles got into the bottle during the night, only to be killed by the sun the following day. These then were bait for the next group. More killed, more bait. There was an accumulation of years of beetles in one bottle, along with a few dead ants, spiders and other creatures. The ordinary insect-eating scavengers would not have any better chance than the first ones in.

TARANTULA HAWKS

(Pepsis chrysothemis)

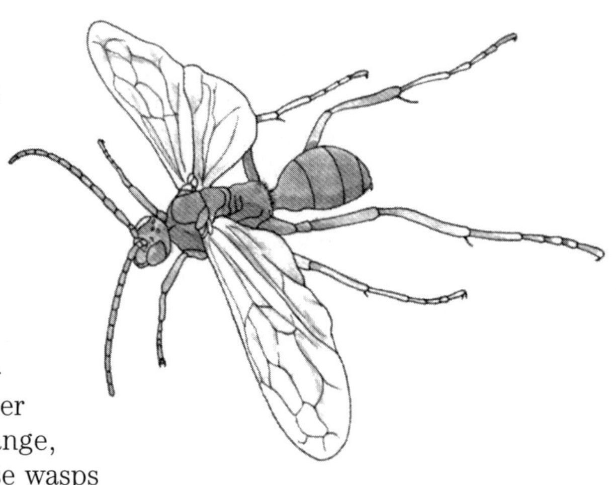

Tarantula Hawk
About actual size.

Just about the largest wasps anyone will ever see are species of tarantula hawks that are found in most parts of the Southwest. The body may be over two inches long, robust and metallic blue-black. The wings are either blue-black or bright orange, the antennae black. These wasps are most often seen around blossoms during the summer, where they are picking up a little fuel for their activities. Mating also takes place there.

The larvae of tarantula hawks are helpless grubs, which depend on their mothers to provide them with a paralyzed tarantula or other large spider. This takes a bit of doing, because tarantula hawks are diurnal and tarantulas nocturnal. The female wasps go into a searching mode running wildly over the ground, using mainly their sense of smell to find the burrow of a tarantula. There they may be able to induce the spider to come out by vibrating the silk strands at the entrance. The tarantula hopefully regards the wasp as potential prey. If the tarantula emerges from her burrow, the wasp maneuvers into position and delivers a sting into the nerve ganglia in the underside of the spider s cephalothorax.

After the capture and paralyzation of the tarantula, the wasp still has a large job ahead of it. She has dug a hole somewhere in the area as a nest site. Now she must return to her hole and drag this very large tarantula to it. The process usually takes place at dusk or nighttime and is rarely seen. At the hole she pulls the tarantula in, deposits a single egg on it, and closes that nest.

The tarantula has only been paralyzed, and lives long enough for the wasp larva to devour it before it dries up, starting with non-essential parts. The larva completes its development by spinning a

silken cocoon and transforming into a pupa. All this happens within inches of this once formidable opponent that is now just a shell of its former self. Paralyzed spiders stolen from the wasps live for months barely able to move, but probably never recover unless meticulously cared for in a laboratory.

Tarantula hawks are the largest representatives of the spider-wasp family Pompilidae. This whole group does essentially the same thing, selecting certain kinds of spiders and making special kinds of nests. The smallest of these wasps are barely as large as a leaf-cutter ant.

VELVET ANTS

(Family Mutillidae)

Velvet Ants
Red & black ant above, white ant below; Actual size.

Velvet ants come in all sizes, but most are from half an inch to an inch long, clothed in long, brightly colored hairs. The commonest colors are orange and black. One spectacular mutillid has erect silvery white hair and looks more like a blowing creosote fruit than an insect.

These are not ants but *wasps.* The individuals most often seen are wingless female wasps that live at the expense of other kinds of solitary wasps. They find the burrows of other wasps by searching the ground and depositing their own eggs in the cells of the rightful owner. Their bodies are hard as stone; they must be able to tolerate any stinger that another wasp might try to stab into them. Being wasps, these beautiful creatures are armed with a most impressive stinger, so beware. Despite this seemingly precarious way of making a living, there always seem to be velvet ants around. The males are less often seen. The diurnal species usually have dark wings that set off their brilliantly colored bodies, and they spend much of their time at blossoms. Males, of course, cannot sting.

The males are so different from the females that the two sexes have not all been matched by entomologists. One way this can be done is to carry around caged females and capture any males that are attracted to the sex pheromones emitted by the females.

A more daring pastime, which I have never had the nerve to try, is to pick up a female velvet ant and spin it rapidly between thumb and forefinger, stopping when the tip of her abdomen is on top of the thumbnail, letting the long stinger flick out over the nail. The stinger, though, may be longer than most people s nails which then certainly results in rapid release and probably a bit of howling, for the sting of a velvet ant is one of the more painful experiences of the bug collector. This tricky display is normally credited, provincially speaking, to our good friends in California whom we love to poke fun at, good-naturedly of course. All of our species have the body so flexible that the female is dangerous when picked up by the thorax.

VELVET MITES

(Order Acari)
(Family Trombidiidae)

One of the characteristic animals of the summer rainy season is the velvet mite, bright orange to red, sometimes with cream markings, and a third of an inch long. They appear on the surface after the soil has been thoroughly soaked. They are quite attractive.

The appearance of these mites is almost certainly coinciding with the great flights of winged ants and termites. The mites are predators, and find easy prey at that time. They can be very abundant. An entomologist visiting Tucson noticed that the desert near Picacho Peak, a mountain between Tucson and Casa Grande, Arizona, was tinged pink when he flew over. When he went back to investigate, he found that the color was of velvet mites.

Not a great deal is known about the rest of the year in the mite s life. Immatures have been found in the egg pods of grasshoppers, so they may be general predators.

Velvet Mite
Twice actual size.

ARMY CUTWORM MOTHS

(Euoxoa auxiliaris)

Owners of mountain cabins very often find that they are inundated with dingy brown moths when they open their cabins in the summer. Astronomers curse them in their mountain-top observatories. The moths almost always turn out to be army cutworm moths, a bit

Flashy in the Mountains

Some insects are either so colorful or numerous that they are noticed by all. There are less of these in the mountains than on the deserts, but their numbers can be overwhelming. The convergent lady beetle could have been included in the hordes of bugs category, but is placed here for convenience. Rocks, trees and everything else can be covered by these insects in the summer.

The galls on oaks form a special category, because they are plant products that result from insect activity.

more than an inch long and two inches in wingspan. Their numbers vary from year to year.

The caterpillars that go with these moths are not found in the vicinity of the cabins, but in the lowlands. They can become pests sometimes, as might be guessed from the common name, but most feed on weeds in wastelands next to agriculture and even out in the desert. The variation in the number of moths clearly correlates with the plant growth in the lowlands that is associated with winter rains. The moths have one to several generations in the lowlands before the adults leave for their vacation cabins in the early summer. They may be arriving when the folks who think that these are their cabins do.

The moths are extremely adept at finding cracks to enter. They actually aren t looking for cabins, but cracks in bark and such places. These cracks are closed in back. The cracks in a cabin aren t or the moths wouldn t be inside. The obvious solution is for the human owner to spend part of his vacation caulking.

GREEN BEETLES

(Family Scarabaeidae)

Four scarab beetles of the genus Plusiotis, famous for their brilliant-green colors, are found in the Southwest. The first three to be discussed come to lights at night and are favorites of beetle collectors. The most brilliant and favored is *Plusiotis gloriosa,* an inch-long

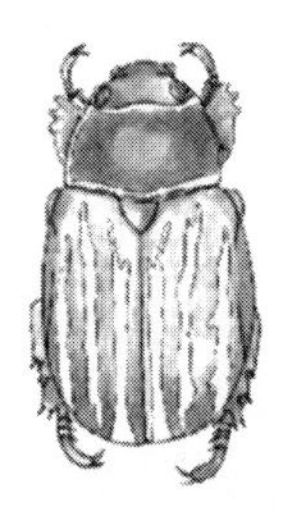
Silver-striped Scarab
Actual size.

character, green with metallic silver stripes on the elytra, hence the common name *silver-striped scarab. P. beyeri* is a matte green with pale-purple legs, *P. lecontei* is a richer green and smallest of these beetles, and *P. woodi* from SW Texas is similar to P. beyeri but has green legs with metallic blue feet and is a daytime flyer.

Despite their brilliant colors, these insects are rarely seen during the day. *Gloriosa* has been seen swarming on juniper trees, perhaps more for mating than for eating. The others just disappear. The larvae of all these species feed on decayed wood, particularly in the underground parts of dead trees or stumps.

Drs. Ross H. Arnett and Richard L. Jacques (1981) made the amazing observation that *P. gloriosa* is becoming endangered, perhaps getting carried along in the endangered species foolishness that seems to be pervading the country. An insect that depends on dead trees and junipers for its existence could hardly run out of habitat in the Southwest U.S., to say nothing of the Sierra Madre Occidental. It is hard to imagine such motile animals as insects becoming endangered by any of mankind s collecting activities, inhabitants of caves and springs being a notable exception. The expansion of the Endangered Species Act to include distinguishable populations of subspecies as well as species has made it possible to declare several insects endangered, mainly butterflies that depend on isolated populations of violets in mountain meadows.

HORNED BEETLES
(Family Scarabaeidae)

Heavy-bodied beetles a couple of inches long and bearing horns on head and prothorax are called *rhinoceros beetles.* There are at least three very different species in the Southwest. The largest is khaki gray, with brown spots, varying to all brown. This is *Dynastes granti.* Males have a horn projecting forward from the top of the prothorax that opposes a horn from the top of the head. Females lack the horns. These beetles are great favorites with

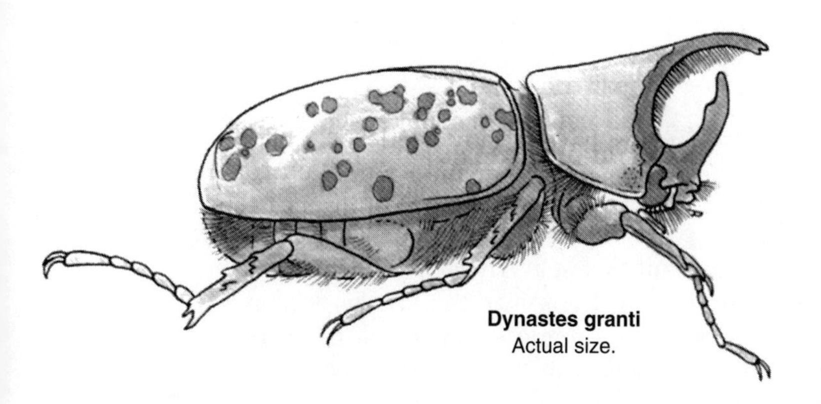

Dynastes granti
Actual size.

collectors, who may make long trips to the Southwest just for these beetles. Large males with long horns are particularly favored. They come to lights and are easy to pick up at street lights. Entomologists are notoriously tight-fisted, but they must at least buy gasoline when they come for beetles.

Strategus julianus is all brown and two inches long. The male has the prothorax dished out and projected into three horns on front and sides. This species may be found at lower elevations than the others. At one time the student collections at the University of Arizona always contained this species because there was a breeding colony in the damp sawdust used in the pole-vault pit.

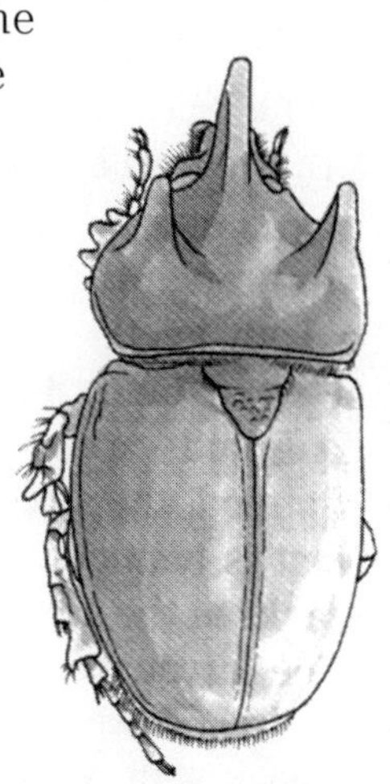

Strategus julianus
Actual size.

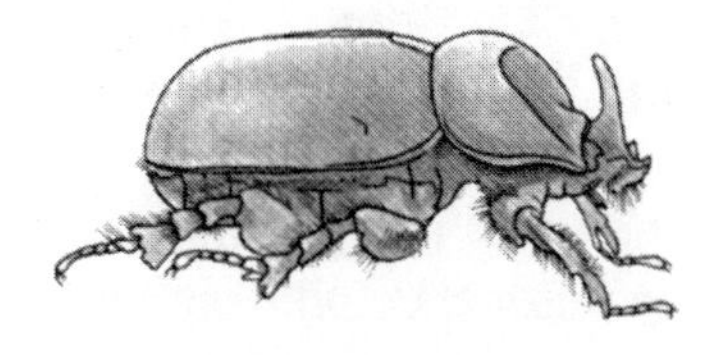

Xyloryctes jamaicensis
Actual size.

Xyloryctes jamaicensis is the smallest of the lot, a little more than an inch long and dark brown. The male has a sharp horn emanating from the top of its head. It is apparently the most abundant and widespread species.

LADY BEETLES

(Hippodamia convergens)

The familiar picture hikers have of lady beetles is a mass of orange beetles with black spots, seen either on a mountain top or in the Sunday supplement when biological control of insect pests is being promoted. The long-handled name for this species is *convergent lady beetle* or the scientific name *Hippodamia convergens.*

Although you may not think it, there are both male and female lady beetles. The common name is a translation of the scientific name, referring to a pair of white convergent dashes on the prothorax, the round plate behind the head. If the namer had looked from the other way, it would have been divergens.

There are many other lady beetle species here, but this is the commonest one. As the biological control articles say, the main food of both the adults and the strikingly colored larvae is aphids. The big aphid season on crops is springtime, although a few species thrive through the summer. For the home gardener, spring is aphid time. Lady beetles may be in evidence, but neither they nor the parasitic wasps and aphid lions that hit aphids really hold the aphids back. Aphids are real breeding machines. Then suddenly, over a brief period, the aphid problem is gone. If one is keeping close track of what is happening, it is at that time that lady beetle larvae join the ranks of the adults in a big way.

The larvae are orange-and-black beaded monsters with sickle-shaped mandibles. They look like a miniature gila monster to many. After feasting on aphids they will form an orange pupal case attached to a leaf somewhere near the feeding frenzy. The adults appear about a week later, usually as the plants dry and the aphids are mostly gone. They depart for the mountains where they feed on pollen and spend the rest of the year just hanging out. If you arrive at one of these sites when the lady beetles first arrive in early May, don't be surprised to have them land on your arm and start biting. They maintain their aggressive feeding behavior for about a week until they discover the best resort area and settle down for the year. These same adults will return to the valleys next spring to help deter the aphids once again.

Nothing is simpler than rounding them up by the bushel and bringing them back down. There is a thriving micro-industry to do just that for a price. However, there is one small picky problem. The

beetles on the mountain are in diapause, a condition in which they feed little and certainly do not mate. Brought down from the mountain they may eat a bit more, but more likely they will leave and head for the hills once again. Much money has been spent by the University of California to find a feasible way to break this diapause, with no commercially viable results so far. So moving the beetles is a waste of money, even if every one stays in your yard, which they won t.

This bright-orange beetle, with black-and-white markings, would seem to be an easy mark for mountain birds and lizards, but the colors are a fine example of warning coloration. All stages of lady beetles have a strong odor, which translates into a very bad flavor when one is eaten in a salad.

Some effort is being made to add a European lady beetle that does not go into summer diapause to the North American fauna, but the species hasn t established itself here yet.

There are many other lady beetles in the Southwest, some blue or blue-black with orange spots. These tend to feed on scale insects. Others are not so brightly colored but have an equally voracious appetite for small plant feeders such as whiteflies.

With all the kind words and praise heaped onto lady beetles, a black sheep does occur in this family. It has turned its back on the good deeds of predation and feasts instead on plants. The Mexican Bean Beetle can raise havoc in gardens, but who would guess it to be kin to normal lady beetles?

OAK GALLS

(Family Cynipidae)

Oak Gall
Actual size.

In yet another case where the things that bugs do live after them, galls on oak trees can be very conspicuous, and the small wasps that cause them completely unknown. All of our species of oaks have some kind of galls. The most obvious ones look like large golf balls and can be tan or tan striped with pink. Some of these have a small cell in the center suspended from threads of tissue, others are almost solid. Other galls look like red wool and still others, nipples or irregular growths.

▶▶▶▶▶▶▶▶▶▶▶▶▶▶▶▶▶▶▶ ♦ ◀◀◀◀◀◀◀◀◀◀◀◀◀◀◀◀◀◀◀◀◀

The wasps that cause these galls to be produced are members of the family Cynipidae. The female wasp introduces a substance into the plant at the time that she inserts an egg. This has to be a very specific material, because it is what causes the plant to produce that species of gall. The feeding of the larva is required to keep the gall growing.

To make things even more interesting, most gall wasp species have two forms, one produced in the fall and winter, the other in the spring and summer. The galls produced by the two forms are different from each other. The fall and winter wasps hatch from the galls that are more noticeable, and insert their eggs into buds. Bud galls are usually small and inconspicuous. The wasps that hatch from these in the summer insert their eggs into buds, leaves or twigs, and cause the conspicuous galls to be produced.

There are some other kinds of galls on plants, produced in response to certain flies, mites, aphids, psyllids or some kinds of larvae. The ones on creosote bush, aloes and palo verde are treated elsewhere in this opus.

Crop Insects

A few insect species cause most of the damage to the crops planted in North America, and only a few more to the crops in the rest of the world. Unfortunately, they can cause a stupendous amount of damage, and greatly increase the cost of production through the need to control them. A great many have developed some level of resistance to the standard insecticides, so that every generation is harder to manage than its predecessor. Thus new trends in management are moving away from strictly pesticide use and more towards use of the diverse tools of IMP (Integrated Pest Management). Some of these methods include changing planting and harvesting dates to avoid insects, early plowdown times to eliminate hidden insect stages, good sanitation practices and biological control means. Some biological control practices include the introduction of predators, parasites and insect disease organisms for more natural management of populations of competitor insects.

The species mentioned here are the main ones of concern in various parts of the Southwest where cotton and grains are grown. Millions of dollars are expended annually for their control.

BOLLWORM

(Helicoverpa zea)

Before the appearance of the pink bollworm and the boll weevil, the insect called the *bollworm* was counted as the most serious pest of cotton in the Southwest. Now it is hardly mentioned. The bollworms are larvae of moths, which fly from plant to plant at night, depositing eggs on the lush terminal growth. The caterpillar that hatches from the egg feeds first in the terminal, then works its way down the plant eating flower buds (squares), then bolls (fruit) as it gets older. One caterpillar can take care of the production of a branch for approximately two weeks.

The same caterpillar is the one you look for when you peel back the husk on an ear of corn. It is called the *corn earworm* in the midwest. The moths are more liberal with their eggs when they put them on corn silk. Anybody trying to produce sweet corn is familiar with the need to do something to keep the larvae out of the ear. The simplest thing is to apply a vegetable oil to the silks.

Not too many years after DDT came in as a standard insecticide for cotton, Peru experienced extreme difficulty in controlling their bollworm on cotton. It turned out to be the tobacco budworm, *Heliothis virescens*, a different species also found in the U.S. More recently this same species has developed a high level of resistance to most of the insecticides now used on cotton in this country. The tobacco budworm was a minor pest of roses in Arizona before this development, and rare enough that we collected all of the specimens we found at lights.

The critical difference between the bollworm and the tobacco budworm is that bollworms have quite a flight period before mating. This ensures wide cross-breeding and cuts down the chances of developing an insecticide-resistant race. The larval stages require the use of a hand lens to see the differences in spiny bumps on the body and teeth on the mandibles, features most people would not recognize at a glance.

BOLL WEEVIL

(Anthonomous grandis grandis)

The boll weevil conjures up visions of the Deep South and the agony it caused when it first hit that area. The weevil is a beetle with a long snout out front that bears the chewing mouthparts at the tip. All that boll weevils eat is cotton, and they can be very destructive.

The weevil's first modern appearance as a pest was in Monterrey, in northeastern Mexico, about 1890. It did not take long for it to reach southern Texas. The spread across the cotton belt was slow, depending mainly on the flight ability of the weevils, but the spread was com-

plete. The diversification of crops that it forced caused the town of Enterprise, Alabama to erect a famous statue in its honor, one of the few insect commemorative statues in the world.

The cotton areas from West Texas to California were never invaded. The assumption was that the requirement of having a place to hibernate in the winter would preclude its establishment in arid areas. Southeastern Arizona had a similar insect, the thurberia weevil, *Anthonomous grandis thurberiae,* on wild cotton or thurberia. It never got beyond the fringes of cotton fields next to the homelands of the weevils. The tight bolls of thurberia kept the weevils inside until summer rains so they couldn't get to cotton early and build up their numbers.

About a decade ago a slightly different boll weevil turned up as a pest of cotton in Arizona. I remember having a man from Caborca bring in a sack of bolls with weevils in the 1950s, so the infestation in Sonora probably goes back before then. The Sonoran problem spread here, and resulted in the massive attempt at eradication now being undertaken.

A combination of pheromone traps to find where the weevils are and repeated applications of insecticides is being used. The traps show that the weevils can live through the winter in places as bare of cover as a table top covered with dust. The hibernating weevils move around in the winter, and find a food supply in the buds and blossoms of globe mallow, *Sphaeralcea* spp., relatives of cotton. They can't reproduce on this plant, but it tides them over. Careful examination of the genetic makeup of the thurberia weevils shows that they have been contaminated with genes from this newly important weevil, so there's little likelihood that the genie will go back in the bottle.

The practice of "stubbing" cotton, breaking the plant off and starting the new season from the roots, is very helpful to both boll weevils and pink bollworms. Areas in which it is practiced are almost guaranteed good populations in the spring. The regulatory branch of entomology has its work cut out for it.

GRASSHOPPERS

(Order Orthoptera)

Grasshopper outbreaks are part of life for many people who live in parts of the Southwest that have good grass cover. Grasshoppers are

Band-winged Grasshopper
1.5 times actual size.

primarily a range problem, but this is no solace to home gardeners who see their entire planting get consumed. The Southwest has a large grasshopper fauna, over 200 species, most of which feed on plants that everybody considers weeds. A few species cause all the trouble.

Unless one lives very close to the breeding area, grasshoppers are likely to leave the garden alone for most of the season, moving in late summer or fall. This is because most of the common species take the whole season to complete their development and become airborne. Females lay their eggs in very specific kinds of places, usually on a gentle hillside with isolated clumps of grass. The hoppers that hatch in the spring depend on a good local supply of food. If the season has been a dry one, they may never make it to the first molt. On the other hand, a rain at the time they are climbing out of the ground can drown the lot. Grasshopper surveys are based on the number of adults in the fall, fine-tuned by a second survey in the spring after the eggs have hatched.

Grasshopper control on rangeland has involved the use of many techniques. In pioneer days one of the methods was the use of a hopper-dozer, a wheeled device drawn across the field that destroyed grasshoppers that jumped up onto a vertical sheet. Then came poison bait, bran with molasses and a poison. Later insecticides were applied by air. DDT, one of the least expensive insecticides for many years, was the logical choice while it was legal. Now ultra-low-volume application of malathion is used most frequently. Federally supported grasshopper control has been based on equal share payment by the U.S., the state and the rancher.

Not everybody wants to get rid of grasshoppers. Many birds, including quail, are dependent on them as an important food supply for the young, particularly while the grasshoppers are young. Natural control by predators and naturally occurring diseases keeps the numbers low most years. Range management to discourage the open spaces that breed annual grasses and weeds at the favored egg-laying sites seems to be the most effective long-term approach. Workers at Montana State University have been working with ways

to get control by spraying with disease-causing organisms like *Nosema,* but this method is still being perfected.

The locust swarms of all of the continents except North America and Antarctica are aerial migrations of grasshoppers. North America once had such outbreaks caused by the Rocky Mountain locust. A commission sent west to study the problem predicted that the fencing of the range and prevention of overgrazing of the favored egg-laying sites in the foothills of the Rockies would take care of the problem. It did. The only Rocky Mountain locusts still found are frozen into Grasshopper Glacier.

PINK BOLLWORM

(Pectinophora gossypiella)

The pink bollworm is an Eastern Hemisphere contribution to the pests on domestic cotton, and was introduced into Texas in 1917. It happens to do much better in the irrigated cotton-growing regions of the West than it does in the Southeast. In fact, it has not spread east of Texas. The adult moths that go with the pink caterpillars in the bolls are small and not particularly good flyers. It was possible to contain the infestation for many years by the simple process of preventing the introduction of bolls and gin trash containing larvae into Arizona. The host range of the species is very narrow, being confined to planted cotton, wild cotton and okra.

For some years an infested area in Greenlee and Graham counties in eastern Arizona was maintained as the western limit of the infestation. Then there was a rapid spread westward to include all of the cotton-growing areas of Arizona and the lower deserts of California. The present situation is that the species has not established itself in the Central Valley of California and the USDA is carrying on a massive program of raising the species, sterilizing the pupae and releasing the sterilized moths in all places in California where moths turn up.

This has been one of the most stubborn species to control with pesticides, because the larvae spend all of their lives in protected places in the plant. The female moth lays her eggs under bracts. When the eggs hatch the larvae have merely to eat into the square (flower bud) or boll (fruit) and they are in their food supply.

Frequent applications of pesticides target the adult moths. Much effort has gone into a program of cultural control devised around removing the cotton plants from the field as early in the winter as possible, to reduce the number of larvae going into diapause. Studies have shown that the percentage going into diapause starts at zero about September 1, and increases with each day that passes. Diapausing caterpillars survive the winter; non-diapausing caterpillars would require a continuous food supply and do not survive the winter. The surviving caterpillars pupate and emerge as moths in the spring, over a considerable period of time. Delaying planting as long as possible leaves the early-emerging moths no place to lay their eggs. Destroying the larvae by plowing and deep burial of plant debris also helps. But the costs of a reduced growing season and the logistics of getting all of the crop picked in good time always keep the program from reaching its ideal.

SPOTTED ALFALFA APHID

(Therioaphis maculata)

When I arrived in Arizona in 1954 the two insects that were causing serious problems were the spotted alfalfa aphid and the khapra beetle. The spotted alfalfa aphid had hit the scene suddenly, so suddenly that it didn't have a common name. It looked like the yellow clover aphid, *Therioaphis trifolii,* another species found on sweet clovers. It took a year to find out where it had come from. When it hit alfalfa in North America, it spread rapidly across the continent, killing alfalfa wherever it went. Infested fields became so sticky with honeydew that insect nets could hardly be used to sample them. Balers got gummed up so badly they wouldn't work and several spontaneous fires were reported where green alfalfa with the high concentration of sugar had been put into grain elevators.

A massive effort to develop a resistant variety of alfalfa started right away, and was successful within two years. Lahontan was the result. Later the aphids overcame Lahontan alfalfa and other varieties came in. We are now on our fourth resistant biotype of the aphid. Insecticidal control was the first line of defense, but the material put on had to break down rapidly enough to be absent at haying time. Parathion worked well, but was too dangerous to apply by ground.

In the News

This last category is the one that probably causes the most concern to people and creates the myths entomologists are constantly working to overcome. Insect invasions are great headline material for papers. The grocery-store tabloids create giant bugs that make us laugh, but the general public may believe such atrocities. Headlines scream about killer bees, but fail to give the true story. Fire ants kill man in Florida, but if you don't read beyond the headline you miss why he was stung to death. Dangerous insects get good press. Our crops are under dire threat from imported pests. To an extent these predictions have some basis in fact. Just don't get overexcited. Nature has a way of balancing the books if given the opportunity.

Malathion became the material of choice. It worked, but generated another pest in the form of leaf-mining maggots, the larvae of tiny flies. The malathion had killed all of their parasites. It took several years to discover cause and effect and come up with a better answer.

Meanwhile, the biological control people in California headed out to do their work. The late Robert van den Bosch found the most effective parasitic wasp, *Trioxys complanatus*, on his first day in Israel, where the aphid feeds on alfalfa and bur clover, as it does here. The parasites are now part of our scene. They and resistant varieties keep the aphid pretty well in check.

ASIAN TIGER MOSQUITO

(Aedes albopictus)

One of the prerequisites of a good control program is to give the object a catchy name. The Asian tiger mosquito is not really very tiger-like, but it does have a brilliant white stripe on its thorax, white bands on the legs and is a ferocious day-time blood-sucker.

This latter behavior led to it being called the *forest day mosquito* in Hawaii, not quite a headline grabber as names go. This mosquito's origin is temperate Asia, but it is now found in Europe, Africa and North and South America.

This species is notorious because of its medical importance to human beings. It is an efficient vector of the Eastern Equine Encephalitis (EEE) and dengue (break bone fever) viruses. Laboratory work has shown it to be capable of vectoring the St. Louis Equine Encephalitis (SLEE) and LaCrosse viruses.

The most likely method of importation and dispersal of this mosquito is the movement of old tires that were stored out of doors, although artificial containers like flower urns or bromeliads provide excellent breeding spots. Rain water impounded in old tires is almost impossible to eliminate. Even if the water is removed, the eggs of the mosquito were probably laid above water line in the tires, would resist dessication and would not be detected by most inspections.

On August 2, 1985, this mosquito was found for the first time in Houston, Texas. This discovery corollated with a shipment of old tires from northern Asia to Houston. Later discoveries of this mosquito in Houston were associated with old-tire dump sites, too. By 1992, populations were found as far north as Nebraska and east to Maryland, with all other states in this region having them present. These mosquitoes also were found in California. Because of the aggressive nature of the larvae of this species, it out-competes most other similar mosquito larvae for habitat space, thus making them a dominant species to be dealt with in the United States today.

CALIFORNIA RED SCALE

(Aonidiella aurantii)

One of the main reasons that oranges have been inspected at Arizona's western borders is that California has long been infested with the California red scale, an insect imported from China, the home base of citrus fruits. The insect appears as small translucent scales on the fruit. There is fear that the young that hatch from these scales will infest trees planted nearby. So all fruit, especially any that has not come from commercial sources, must be inspected. The program has never been popular with motorists and the "bug stations" have received negative attention from the legislature almost every session.

A potent reason for avoiding getting the scale is that Arizona happens to be one of the few places in the world where citrus fruit can be grown without application of insecticides. Japan, in particular, has become very aggressive about banning the importation of anything with any possible toxic residue on it. So Arizona citrus exports at a premium price.

The importation of a parasitic wasp, the golden-eyed chalcid, *Aphytis melinus*, from China has reduced the scale populations in California to the level that they do not really affect production. It took several importations, because citrus grows in several different climatic regions. Now there is a full-scale rearing-and-release program to keep this scale under control.

GYPSY MOTH

(Lymantria dispar)

We owe this pest, known as the *dancing poison moth* in China, to one Leopold Trouvelot, a French astronomer who was going to do American industry a big favor when he imported it from Europe in 1869 to cross with the silk worm. Leopold must have been a better astronomer than a geneticist. The moth escaped from his cages and soon established itself as one of the worst pests of woody plants this continent has ever seen. The moth itself is a drab one and the female flightless. The rapid spread of the species came as the result of young caterpillars letting out a strand of silk and "ballooning" along on the wind.

Movement beyond the ballooning usually comes from the tendency of the caterpillars to make cocoons in sheltered places, including the undersides of trucks and trailers. Campers moving around infested areas are almost sure to pick up a few. Attempts at control during the 20th Century have ranged from placing a bounty on cocoons and egg masses to widespread application of insecticides.

The infested area has been confined mostly to the northeastern United States with isolated populations established in California, Oregon and Washington. There has been a large investment in finding and importing parasites and predators to try to keep the populations

in check, hoping to reduce the chemical warfare used against this moth. These tactics have helped, but an outbreak and massive defoliation occurs about once a decade.

From the extremely low populations that normally occur, it takes the gypsy moth about two years to escalate its numbers to horror-movie size, but within 2-4 years the population will again collapse. This collapse remained a mystery until the 1900s when researchers at Penn State University discovered that trees like the red oak fought back against these defoliators by releasing phenols, defensive chemicals, into their leaves. The phenols stunted the growth of the caterpillars and reduced the fecundity (egg production) of the female moths, resulting in a dramatic population crash. Many other trees that the gypsy moth feeds on, including aspen, alder, willow, maple and apple also have "induced defensive" responses that help curb the moth populations. Much more is being discovered about bacterial, fungal and viral diseases of the gypsy moth and how the trees' normal defenses affect these controls. Nature does seem to have a lot of answers for overpopulation problems, given time to set them into motion.

A recent extremely wet spring in the Northeast resulted in a poor gypsy-moth population. It was discovered that the brown-tailed fungus, an imported control for gypsy moth had never really worked because of poor climatic conditions. With the heavy soaking that occurred, the fungus population that was hanging on suddenly burst forth, and flaccid gypsy-moth larvae, infected by the fungus, were found hanging from the trees, dead.

Another species of gypsy moth, this from Siberia, has recently been found in Washington. Although panic has not ensued in the lumber industry yet, this species could be more insidious because the female moth can fly, thus adding potentially rapid spread throughout the forests.

HONEY BEE/AFRICANIZED BEE

(Apis mellifera)

The European strain of honey bee, *Apis mellifera mellifera,* is familiar to all. It is about the only bee one is likely to see in abundance except in the most remote locations. Its dark-honey color can vary a

bit, some strains being quite dark on the abdomen. Colonies are maintained in hives that permit easy management and removal of honey and comb as commercial products. Maintained colonies are provided for pollination of several crop plants. New colonies are formed by the swarming of many workers with a single queen, usually the queen of the original colony. Part of good management consists of preventing swarming.

Honey Bee
Actual size.

Feral colonies can establish themselves by swarming and maintain themselves for many years. These are usually in cavities, but occasional colonies set themselves up out-of-doors, suspended from a branch. Research has shown that feral colonies like these are very aggressive if they have brood, so be careful.

Honeybees are the only stinging insects in North America that have the stinger barbed so that it cannot be withdrawn. When the bee pulls away from the sting site, the stinger remains behind, along with the reservoir of venom and muscles that continue pumping. Removal by scraping with a knife prevents squeezing more venom into the wound. Some people have become very highly sensitized to the sting and must carry medication to prevent drastic symptoms.

A great deal of publicity has been given during the past 20 years to the Africanized (or so-called *killer*) bees. Africanized bees look exactly like the honeybee everyone recognizes because it is simply a subspecies, *Apis mellifera scutellata.* They are a bit smaller but it takes much dissection work and a sophisticated computer program dealing with measurements of different structures and ratios of these measurements to tell the difference accurately.

Honey bees have been domesticated for a long time in Europe and North Africa. During this long period early beekeepers bred into them a number of traits, including docility, a trait probably suppressed in the Africanized strain. Ordinary domesticated bees do not do well in the Tropics, so a Brazilian worker attempted an experiment to determine whether wild bees from Africa bred with South American domesticated stock would improve bee survival in that country's tropical zone. Some of the imported swarms escaped and have spread northward, inhabiting Central America, Mexico and most recently southern Texas and Arizona. The hoped-for outbreeding of aggressiveness has not occurred, but improved survival in the tropics has.

Typical behaviors of the Africanized bee include increased swarming, up to seven times during a typical season. This behavior has helped in the rapid spread of this bee. The aggressiveness may be a bit exaggerated, but these bees do respond quickly to any disturbance of their brood area, including loud noises like lawn mowers, fast movements nearby or scents like perfumes or cologne. Certainly mothers of any animal are quick to defend their offspring so these ladies can't be blamed for this behavior.

Bee keepers in South America have learned to keep their hives widely separated so no inadvertent bumps alarm the other colonies. The sting is no more potent than the European strain of honey bee, but usually more workers are drawn into the attack, so more stings occur. Even with all these potential dangers, these beekeepers may still tend their hives sans veils many times.

Outdoor activities like hiking and climbing need not be abandoned due to this new situation, but common sense must be used. If one encounters a swarm of bees, the best scenario is to leave quietly. If the bees become agitated and come towards you, run and seek a shelter inside a car or building. Do not start hitting at the bees, just get away. Usually the bees will not follow more than half a mile.

In urban areas, swarms are routinely encountered and a swarm of bees sitting on a tree in your yard really poses no threat. One does not want these bees to establish a hive nearby, so it is wise to screen vents and openings into your attic or walls. If a tree in your yard or nearby has an opening, it is wise to plug that up too. If bees do establish a hive, call your fire department or state agriculture department. They have the training or can recommend someone to remove the bees. ***Do not try to remove them yourself.*** Again common sense must come into play when dealing with these animals. The ordinary homeowner wouldn't try to get rid of a bear that wandered onto the property. Bees can create just as threatening a problem. Call in the experts.

If the dire predictions of the popular press were at all accurate, it would be hard to explain the burgeoning human population in Africa and Brazil. It will certainly be more demanding to maintain colonies, particularly in urban areas, but beekeeping is too important to crop pollination to be abandoned. With proper management of domesticated honey bees, problems of Africanization of colonies can be prevented or readily dealt with. A healthy hive will resist unwanted takeovers.

Honey bees were brought to the Western Hemisphere by European colonists. Left behind were at least part of the pests and diseases that affect them. Tracheal mites, which live inside the tracheae of the respiratory system, have recently made the trip across from Europe, probably in some illegally imported bees. Infested colonies become unthrifty and may even die out. Apiarists are at least as concerned about these mites as they are about their colonies becoming Africanized.

IMPORTED FIRE ANT

(Solenopsis invicta)

Most of the serious insect pests that we get from other parts of the world come from Eurasia, the original home of many of our crops. The imported fire ant is an exception, hailing from temperate South America. It is a stinging ant, and would be hard to distinguish from the southern fire ant that is native to the Southwest. It is all red, and much more aggressive. In the Southeast where it is a problem and where much money has been spent trying to eradicate it, the large mounds can be a familiar sight. Farmers get stung when their equipment hits the mounds, sprouting seeds are consumed, young birds eaten, etc. It is a hard animal to get along with.

The Southwest has mostly escaped invasion, although the steady importation of all kinds of products from the East must inevitably establish it here. Nursery plants in pots are probably the most likely source, although it is possible that a mated queen hidden away in a truck could be the pioneer we can live without. Infestations have been discovered in the Phoenix, AZ area and southern California in recent years, but responses by the state Agriculture Departments have kept them in check.

The most likely point of importation for North America was Mobile, Alabama, and the most likely source, ballast from Argentina. Colonies of a black fire ant became common in the Mobile region but the ant did not expand its range until some red ones began turning up. The current feeling is that these represented a second importation. They quickly supplanted the black imported fire ants and began increasing their range.

The ants are not an undivided curse. They are very vigorous predators, and must be very beneficial in keeping insect pests in check. Few vote in their favor, however, especially now that eradication efforts have been abandoned. The dire predictions about what would happen have turned out to be exaggerated, but the insect is not easy to live with.

Another species from South America, the non-stinging Argentine ant, *Iridomyrmex humilis,* is about as much of a pest. It establishes colonies that overrun everything, the workers becoming pests in the home as well as in the garden, where they tend aphids that feed on plant sap. Southern California has these and they are among the insects that prevent one from bringing potted plants to Arizona without having been through quarantine inspection.

JAPANESE BEETLE
(Popillia japonica)

The Japanese beetle is a small scarab beetle, shiny green with metallic-brown elytra, which became established in eastern North America earlier in this century. It has become a regular pest of lawns and flowers there, the very symbols of the good gardener. The adults are very partial to blossoms, especially roses, and the larvae are white grubs that eat the roots of grasses.

When the species was introduced into New Jersey, its name and home country were not known. In fact, it is not a major pest in either Japan or China, where it is native, so there was no particular reason to be watching out for it. Several parasites have been imported for reducing its populations, but there still are good numbers most years. A bacterial disease, milky disease, has been quite effective in controlling larvae. Once the adults are out, special traps baited with attractant are the most efficient way of cutting down numbers. Such traps are a common sight around airports in areas where the beetle has not yet become established. They are maintained as a check on the possible importation of adults. Ohio had quarantined the eastern half of the state in the 1960s, but the beetle failed to obey the laws and continued spreading westward. Funny how most insects blatantly disregard people rules.

LYME DISEASE TICK

(Ixodes pacificus)

"When you pick up a trap, wherever it's been set, you may be picking up more than a dead mouse. There may also be disease-carrying ticks. And fleas. And mites. These parasites can carry serious diseases such as Rocky Mountain Spotted Fever, Colorado Tick Fever, Tularemia, Typhus and Hanta virus."

"Recently, you may have read about Lyme-disease ticks carried by the white-footed field mouse. People bitten by the Lyme-disease tick can suffer temporary paralysis of the facial nerves, pain in the joints, and even severe neurological symptoms similar to multiple sclerosis." Tucson Daily Citizen, Jan. 13, 1990! d-CON® advertisement.

With scare items like this in the advertising world, it is little wonder that some people are paranoid about mice, rats, parasites and disease. From our standpoint in the Desert Southwest, the chance of being infected with any of the listed diseases is very slight. The possible exceptions are tularemia, which can be picked up through cuts when dressing a sick rabbit (a rare disease indeed), and Hanta virus. The quoted ad was for killing domestic mice and rats, *not* wild rodents.

In the case of Lyme disease, which continues to receive national publicity, the ticks that have been proven as vectors are deer ticks, three-host hard ticks that are in no way even similar to the mites that occasionally are found on house mice. In the eastern U.S. and Pacific Coast, where Lyme disease has been found, field mice do indeed serve as hosts for hatchling deer ticks, larger mammals for the two later stages. Even there, though, field mice are not usual household invaders, especially during tick season.

Only one of the species of ticks that have been implicated as vectors of Lyme disease anywhere else in the country has ever been found in Arizona, and this one thus far only in the Hualapai Mts. It is readily found on the coast of California to British Columbia. Small populations also have been recorded in Idaho, Nevada and Utah. Actual human bites by any kind of tick are almost unheard of in desert regions, but in the moister mountain canyons and wooded areas, potential tick attacks may occur in the Southwest.

MEDITERRANEAN FRUIT FLY
(Ceratitis capitatus)

This is a rather pretty little fly, with picture wings (banding with yellow, brown and black) and a bad reputation. It is about a quarter inch long, and yellow except for black markings on the thorax and white bands on the abdomen. It is quite unrelated to the small tan flies that get on ripe fruit during the summer.

It belongs to a fairly large group of flies that feed on ripe fruit in the maggot stage. All are very destructive. Not all of them can infest all kinds of fruit, but the Mediterranean fruit fly, *Medfly* as denoted in print, comes close. They cause great destruction where they occur, including the Hawaiian Islands.

The first time the Medfly was introduced into North America was in Florida in the 1920s. It was eradicated by a Federal/State effort in which infested fruit was buried and all fruit leaving the area inspected for infestation. This was a costly effort, because it involved such tremendous manpower. Since then the fly has gotten into the country at least twice in Florida and twice in California. The advent of modern pesticides has made chemical control the choice, most recently by helicopter application over the Los Angeles area. Fly populations are reduced with malathion, and finally knocked out by release of sterile flies, as in the screwworm program.

Careful scrutiny of fruits is needed to keep fruit flies out. One may not bring mangoes in from Mexico, for instance. Pineapples are passed because the larvae of fruit flies cannot complete their development in them.

PLAGUE FLEAS
(Order Siphonaptera)

There isn't a single species of flea that is called the *plague flea,* but fleas are involved in transmitting the disease from wild rodents to man. The oriental rat flea, *Xenopsylla cheopis,* is the principal vector of bubonic plague. The rodents most likely to be involved in this region are prairie dogs in the northeastern quadrant of Arizona and northwestern New Mexico. Plague is a disease of rodents that is sometimes transmitted to rabbits. People get the disease from bites

Old News

Several of the disasters of yesteryear involved insects. Perhaps the biggest event in Arizona's entomological history was the campaign to eradicate the khapra beetle. Whole city blocks in Phoenix were covered with tarpaulins and fumigated with methyl-bromide gas. Nobody had ever seen anything like it.

The successful effort to eradicate the screwworm was newsworthy more in rural areas. At the peak of the campaign millions of sterile fly puparia were dropped from aircraft every week. The rectangular cardboard boxes became common findings on rangeland.

of fleas that have left dead or dying animals. One of the possible routes of infection is through dogs picking up the fleas in their investigations of the habitat.

Plague is as much a disease of fleas as it is of mammals. A flea that has contracted the bacterial disease gets a blockage of the digestive tract. Without nutrition the flea gets hungrier, but when it feeds it tends to regurgitate because of the blockage.

Great epidemics of plague have devastated whole societies in the past, usually centering around port cities. The home base seems to be central Asia. The last epidemic got to San Francisco in the early 1900s and apparently spread from there into native North American rodents. The way in which the disease spread was through movement of black rats from port to port. The rat fenders on the mooring lines of ships are designed to keep them from getting on or off a ship.

During the past few years evidence has been coming in that the disease has jumped in Arizona to at least the Chiricahua Mountains. The screening method for the disease is to check the blood of coyotes. These animals are not much affected, and pick it up from the rodents and rabbits on which they prey. Public-health authorities keep close track of the disease. They recommend particular caution around prairie-dog towns, pack-rat burrows and when sick rodents are found.

The disease is not as dreaded as it used to be. People do die from it almost every year in Arizona and New Mexico, but it is treatable if diagnosed in time. Public-health workers move in whenever there is indication of plague, and carry out a vigorous campaign of dusting persistent insecticide into the rodent holes.

KHAPRA BEETLE

(Trogoderma granarium)

This insect could also have been treated under dermestid-beetle larvae, but it has had such an impact on Arizona that it warrants being discussed separately. It is a dermestid beetle, with very small, hairy tan larvae, and closely related to several species that are pests of stored grain, animal products, and even stored woolens. At the time of its introduction into North America in the early 1950s, it was the only major pest of stored products we didn't have. Its first appearance in San Francisco was overlooked because nobody really knew how to tell it from the species related to it. George Okumura, of the California Department of Agriculture, found a feature on the underside of the upper lip of the larva that looked like a 4-holed button that distinguished the khapra beetle.

To attack the problem, a massive fumigation effort was employed. The beetles had apparently come in on bagging that had been used to transport grain to the Orient. The bags got distributed from granary to granary, scattering beetles as they went. By the time the problem was discovered, hundreds of properties were infested. Arizona was prime khapra territory. The first line of attack was to eliminate the beetles from the major granaries. In Phoenix this involved covering large grain-storage buildings, several stories tall, with plastic tarps and fumigating them with methyl bromide. The covering and fumigation could take a week or more, during which time no business could be transacted. Obviously, a granary owner who had just had this done would not welcome anybody's sacking that hadn't been treated. The fumigating was free, but the lost business wasn't.

Then the secondary places were sought out, individual buyers who had carried khapra beetles with them on bags of grain, pet food

and the like. These were fumigated. After several years the project was deemed a success. Incoming shipments are now kept under surveillance. A small infestation in a spice factory several years ago was taken care of right away.

The impact of the beetle was impressive. Windrows of larvae several inches thick accumulated on the top of bins, and storage loss of more than 10% was usual. Khapra beetles, like all stored-product insects, create their own environment inside a grain bin, increasing both temperature and humidity. If the khapra beetle had gotten into the major grain-storage areas in the Plains States, spontaneous combustion could have become a common disaster. The cold parts of the stored grain condense the moisture, setting up fermentation, and then look out!

In the Middle East an ancient grain-storage method prevents khapra beetles and other insects from building up. This is underground storage in waterproof dry wells. The respiration of the grain increases the carbon-dioxide content of the air to the point that insects cannot survive in it. People are not that sensitive to carbon dioxide, so can enter the bins with impunity. It happens that insects are very susceptible to this gas. Carbon dioxide is the anesthetic of choice for an operation on an insect.

SCREWWORM

(Cochliomyia hominivorax)

Eradication of the screwworm in North America is one of the success stories of modern entomology. The worm is the maggot of one of the green blow flies, a native of the Western hemisphere, and closely related to another species that differs only in details of larval habits. The screwworm fly deposits eggs next to wounds of living animals. The maggots that hatch from the eggs burrow into the wound to feed, concentrating on the margins of the wound and making it deeper.

Newborn calves and their mothers are very vulnerable, as are dehorned and branded animals. Dogfight participants in Tucson used to be infested, and even drunks who slept it off under a mesquite tree, the maggots making their way up the nasal passages into the sinuses and the brain.

The idea for controlling these insects came from the observation that the female flies mate only once, the males repeatedly. Thus by releasing sterilized males, chances of a female being successfully inseminated were greatly reduced. It later became obvious that one needn't release only sterile males, a burdensome task sexing millions of pupae, so both sexes were sterilized and released. These are irradiated as pupae, put in cold storage, and later added to the environment in the hundreds of millions by air drops.

The first field trial of this method was on a small island in the West Indies, where screwworms were well established in wild goats. Within a year the island of Curaçao was fly-free. Congress was willing to underwrite a more massive program on the continent. In the eastern U.S. the only place the pupae could make it through the winter is southern Florida. A quarantine was set up to prevent infested animals from being brought in from the West, and a fly-release program started for southern Florida. The first problem to be solved was finding a suitable diet for rearing the larvae. Ground beef works, but tends to be a bit costly. A synthetic diet was finally developed and used to generate hundreds of millions of larvae per week.

The eradication was complete for eastern North America within a couple of years. Then a more elaborate program, involving cooperation with the Republic of Mexico, was developed. A fly-free zone would be established across the Southwest, from the Gulf of Mexico to the Pacific, then pushed southward to the narrowest part of the continent at the Isthmus of Tehuantepec. It took establishment of a new fly factory in Texas and a distribution system at scattered airfields across the Southwest. Boxes of fly puparia were dumped on a grid pattern in any area where veterinarians had found a screwworm infestation.

Maggots found in wounds were sent to a center for identification. This part of the program took about ten years to accomplish, but the push to the Isthmus of Tehuantepec went amazingly rapidly. There is now a fly factory at Tuxtla Gutierrez, Chiapas and the program is going strong there. It turned out that success is much more rapid in the Tropics than it was here. Maintenance of the barrier is quite inexpensive in comparison with the costs of the infestation.

The impact on the livestock industry has been major. Not only has there been a reduction in summer loss, but the need to examine the herd for wounds has decreased. The impact on the populations

of wild mammals has been impressive. In the eastern U.S. there are big populations of deer in areas where deer hadn't been seen for generations. Texas is overrun with deer, as are many areas in the Southeast. Removal of such a significant mortality factor has made a lot of difference.

In the late 1980s Libya developed an infestation derived from lambs imported alive from South America. A multimillion-dollar program was undertaken, buying and releasing sterile flies grown in Mexico. Also the veterinary services there were mobilized to help fight the problem. The program seems to have been successful as no reports of animals being infested have been reported in the last few years.

Index